# Lecture Notes in Mathematics

Edited by A. Dold, Heidelberg and B. Eckmann, Zürich



376

David Bridston Osteyee
Office of Naval Research, Pasadena, CA/USA

Irving John Good
Virginia Polytechnic Institute and State University,
Blacksburg, VA/USA

# Information, Weight of Evidence, the Singularity between Probability Measures and Signal Detection

Springer-Verlag
Berlin · Heidelberg · New York 1974

AMS Subject Classifications (1970): 60-G-15, 60-G-30, 60-G-35, 62-F-15, 62-H-15, 62-M-10, 94-02, 94-A-05

ISBN 3-540-06726-4 Springer-Verlag Berlin · Heidelberg · New York
ISBN 0-387-06726-4 Springer-Verlag New York · Heidelberg · Berlin



Offsetdruck: Julius Beltz, Hemsbach/Bergstr.

# PREFACE

The technologies and sciences of information processing and communication have made great advances during the last thirty years and have also stimulated much first-class theory. This monograph contains a survey of the part of this theory dealing with measures of information and evidence and especially with the detection of signals in noise. Although the theory arises directly from practical problems and realistic models, some of it turns out to be surprisingly advanced mathematically.

If all the proofs were included the book would be very long, and we decided that it would be better to omit most of them.

Acknowledgements. The work was supported in part by the Department of Health, Education and Welfare, Grants #ES 00033 and #1 R01 GM18770.

We wish to thank Becky Clevenger for her expert typing.

December 1973

D.B.Osteyee, Pasadena, California
I.J.Good, Blacksburg, Virginia

# TABLE OF CONTENTS

CHAPTER

CHAPTER

CHAPTER

CHAPTER

CHAPTER

CHAPTER

# INTRODUCTION

The present somewhat informal monograph is a study of the circumstances resulting in (i) a probability measure being singular with respect to another one, (ii) infinite expected mutual information, (iii) infinite expected weight of evidence, and (iv) infinite divergence. Several conditions that rule out these situations are also investigated. Most of the results are for gaussian processes in communication theory.

There has been much research done in these areas and we shall survey this research, but, except in Section 15.3.6, we do not discuss the method of "reproducing kernel Hilbert space" for which the reader is referred to Jørsboe (1968).

First, information, mutual information, weight of evidence, entropy, and singular probability measures are defined and discussed. Then the expected mutual information, the expected weight of evidence, and divergence are investigated for random variables, vectors, and random processes over intervals of time. Their relationship to singularity between probability measures and errorless discrimination between two random processes over arbitrarily small intervals of time is then discussed. Several random processes such as ergodic strictly stationary processes with analytic covariance functions, singular processes, and gaussian processes with proportional covariance functions are investigated and related to singularity between probability measures.

The expected mutual information and expected weight of evidence are then expressed in terms of functionals and integral operators so as to include generalized gaussian processes such as white noise.

A comparison is then made between these expressions for the special case of independent stationary gaussian signals and noise. It is observed that the expected weight of evidence may be finite even though the expected mutual information is infinite.

The expected mutual information rate, the channel capacity, and the expected weight of evidence rate are investigated for single and multidimensional random processes, with particular emphasis on independent stationary gaussian signals and noise.

Gaussian processes with equal covariance functions (which include nonrandom signals in random noise) are investigated. The relationship between probability measures and errorless signal detection is also discussed for these processes.

A summary of the major results for gaussian processes, including gaussian signals and noise is given in Chapter XV along with some considerations concerning white noise.

## LIST OF SYMBOLS*

| SYMBOL | DEFINITION | SECTION |
|---|---|---|
| $(\Omega, \mathcal{A}, P)$ | Probability Space | 1.1 |
| $\Omega$ | The sure event | 1.1 |
| $\mathcal{A}$ | $\sigma$-field of events | 1.1 |
| P | Probability | 1.1 |
| A,B,... | Events in $\mathcal{A}$ | 1.1 |
| I(A) | Information about A | 1.1 |
| $\cap$ | Intersection | 1.1 |
| $\Sigma$ | Sum | 1.1 |
| $\Pi$ | Product | 1.1 |
| $\mid$ | Given | 1.2 |
| AB | $A \cap B$ | 1.2 |
| I(A:B) | Mutual information between A and B | 1.3 |
| H | Hypothesis | 1.4.1 |
| $W(H_1/H_2:B)$ | Weight of Evidence | 1.4.1 |
| O | Odds | 1.4.1 |
| $P_H$ | Conditional probability given H | 1.4.2 |
| $W(H_1:B)$ | Weight of Evidence when $H_2 = H_1^c$ | 1.4.3 |
| $H^c$ | Complement of H | 1.4.3 |
| I(X) | Entropy or expected information | 2.1 |
| $p_i$ | $P[X = x_i]$ | 2.1 |
| E(x) | Expected value operation | 2.1 |
| $p_{ij}$ | $P[X = x_i, Y = y_j]$ | 2.1.1 |
| I(X,Y) | Information for the joint distribution | 2.1.1 |

* Intended for reference.

LIST OF SYMBOLS (Continued)

| SYMBOL | DEFINITION | SECTION |
|---|---|---|
| $\mathbf{X}$ | Vector (finite or infinite dimensional) | 2.1.1 |
| $p_j(i)$ | $P[X = x_i \mid Y = y_j]$ | 2.2 |
| $I(X\|Y)$ | $E_{XY}(I(x\|y))$ | 2.2 |
| $f_X(x)$ | Continuous probability density function | 2.3 |
| $h(X)$ | $E_X(-\log f_X(x))$ | 2.3 |
| $h(X\|Y)$ | $E_{X,Y}(-\log f_{X\|Y}(x\|y))$ | 2.4 |
| $\frac{dP_o}{dP_1}$ | Radon-Nikodym derivative of $P_o$ with respect to $P_1$ | 3.1.1 |
| $\int_A dP$ | $P(A)$ | 3.1.1 |
| $P_o \parallel P_1$ | $P_o$ and $P_1$ are parallel (mutually absolutely continuous) | 3.1.1 |
| $P_o \perp P_1$ | There exists a set A such that $P_o(A) = 1$ and $P_1(A) = 0$ | 3.1.3 |
| $F(x)$ | Probability distribution function | 3.2.2 |
| $P_{XY}(A)$ | Joint probability; $P[(X,Y) \text{ in } A]$ | 3.2.3 |
| Min(a,b) | The minimum value of a and b | 3.2.3 |
| $P_1 \times P_2$ | Product probability | 3.3 |
| $F_X \times F_Y$ | Product probability distribution function | 3.3.1 |
| $I(X:Y)$ | Expected mutual information | 4.1.1 |
| sup | Least upper bound | 4.3 & 5.1 |
| T | An ordered set of real numbers such as the time interval $[0,T]$ | 4.3 & 7.1 |
| $\mathbf{X}_T$ | $\{X(t) \mid t \text{ in } T\}$ | 4.3 & 7.1 |
| $P_{\mathbf{X}_T\mathbf{Y}_S}$ | Joint probability measure | 4.3 & 7.1 |
| $P_{\mathbf{X}_T} \times P_{\mathbf{Y}_S}$ | Product probability measure | 4.3 & 7.1 |
| $\|\mathbf{A}\|$ | Determinant of the matrix $\mathbf{A}$ | 4.6 |

LIST OF SYMBOLS (Continued)

| SYMBOL | DEFINITION | SECTION |
|---|---|---|
| $\mathbf{A}$ | Covariance matrix | 4.6 |
| $\rho$ | Correlation coefficient | 4.6 |
| $B_{\mathbf{Y}}$ | Linear space over the components of $\mathbf{Y}$ | 4.6.1 |
| $\hat{X}$ | Perpendicular component of X on $B_{\mathbf{Y}}$ | 4.6.1 |
| $E_1$ | Expected value operator given $H_1$ is true | 5.1 |
| $W(H_1/H_2)$ | Expected weight of evidence given that $H_1$ is true | 5.1 |
| tr | Trace of a matrix | 5.6 |
| $I$ | Identity matrix | 5.6 |
| $J(1,2)$ | Divergence between $H_1$ and $H_2$ | 6.1 |
| $T_n$ | $\{t_i \mid i = 1, \ldots, n\}$ | 7.2 |
| $\mathbf{X}_n$ | $\{X(t_i) \mid i = 1, \ldots, n\}$ | 7.2 |
| $T_n \to T$ | Riemannian limit | 7.2 |
| $\mathbf{X}_{1T}$ | $\{X_1(t) \mid t \text{ in } T\}$ | 7.3 |
| $P_{1T}$ | Probability measure for $\mathbf{X}_{1T}$ | 7.3 |
| $W_T(H_1/H_2)$ | $\int_{\Omega_T} \log \frac{dP_{1T}}{dP_{2T}} dP_{1T}$ | 7.4 |
| $J_T(1,2)$ | Divergence for the processes $\mathbf{X}_{1T}$ and $\mathbf{X}_{2T}$ | 7.5 |
| $m(t)$ | Mean function | 8.1.2 |
| $\gamma(\tau)$ | Covariance function | 8.1.3 |
| $\sigma^2(t)$ | Variance function | 8.1.4 |
| $Y^*(\omega)$ | Fourier transform of $Y(t)$ | 8.2 |
| $\omega$ | Frequency in cycles per unit of time | 8.2 |
| $\lambda$ | Frequency in radians per unit of time | 8.2 |
| $F^*(\omega)$ | Spectral distribution function | 8.2 |
| $dF^*(\omega)$ | Stieltjes differential of $F^*(\omega)$ | 8.2 |
| $P^*(\omega)$ | Continuous spectral density function | 8.2 |

LIST OF SYMBOLS (Continued)

| SYMBOL | DEFINITION | SECTION |
|---|---|---|
| $p^*(\omega_i)$ | Discrete spectral mass function | 8.2 |
| W | Frequency band width | 8.4 |
| $N_W(t)$ | Band limited white noise | 8.4 |
| $\delta(\tau)$ | Dirac delta function | 8.5 |
| $N(\phi)$ | $\int_{-\infty}^{\infty} N(t)\phi(t)dt$, where $\phi(t)$ is a function of t | 8.5 |
| $\Phi = \{\phi\}$ | Set of functions | 8.5 |
| $P_{N_T}$ | Probability defined on N(t), t in T | 8.7.1.2 |
| $\alpha$ | $\sigma_S^2/\sigma_N^2$ | 8.9.2 |
| $\delta_{ij}$ | $\delta_{ij} = \begin{cases} 1 & i = j \\ 0 & i \neq j \end{cases}$ | 9.2.1 |
| $\|\|h\|\|$ | $\sqrt{Eh^2}$ | 9.2.2.1 |
| $H_X$ | Hilbert Space | 9.2.2.1 |
| I | Identity operator | 9.2.2.1 |
| $E_X$ | Projection operator | 9.2.2.1 |
| $\lambda_i^2$ | Eigenvalue of an operator | 9.2.2.2 |
| $L_2(T)$ | $\{f(t) \mid t \text{ in } T \text{ and } \int_T \|f(t)\|^2 dt < \infty\}$ | 9.2.3.3 |
| l.i.m. | Limit in mean square | 9.2.3.3 |
| $L_2$ | $\{f(t) \mid \int_{-\infty}^{\infty} \|f(t)\|^2 dt < \infty\}$ | 9.2.3.3 |
| $B_{XX}$ | Integral operator | 9.2.5 |
| $\mathbf{X}$ | $\{X(t) \mid -\infty < t < \infty\}$ | 12.1 |
| $\overline{I}(X:Y)$ | Information rate | 12.1 |
| $P_{XY}^*(\omega)$ | Cross-spectral density function | 12.2 |
| C | Channel Capacity | 12.6.1 |
| $S_G$ | Gaussian signal | 12.6.1 |
| ${}_n\mathbf{X}(t)$ | $(X_1(t), \ldots, X_n(t))$ | 12.7 |

LIST OF SYMBOLS (Continued)

| SYMBOL | DEFINITION | SECTION |
|---|---|---|
| ${}_n\mathbf{X}$ | $\{{}_n\mathbf{X}(t) \mid -\infty < t < \infty\}$ | 12.7 |
| ${}_n\mathbf{X}_T$ | $\{{}_n\mathbf{X}(t) \mid t \text{ in } T\}$ | 12.7 |
| ${}_n\mathbf{X}_\ell$ | $\{{}_n\mathbf{X}(t_i) \mid (i = 1, 2, \ldots, \ell)\}$ | 12.7 |
| $\gamma_{X_iY_j}(\tau)$ | $E\{X_i(t)Y_j(t + \tau)\}$ | 12.7.1 |
| $P^*_{X_iY_j}(\omega)$ | Fourier transform of $\gamma_{X_iY_j}$ | 12.7.1 |
| $A^*(\omega)$ | Matrix of Spectral density functions | 12.7.1 |
| $\lvert\mathbf{A}^*(\omega)\rvert$ | Principal minor of $\mathbf{A}^*(\omega)$ | 12.7.1 |
| $\overline{W}(H_1/H_2)$ | Rate of Expected Weight of Evidence | 13.1 |
| $X_{1i}(t)$ | ith element of $(X_{11}(t), \ldots, X_{1n}(t))$ | 13.3 |

# I. INFORMATION IN EVENTS AND WEIGHT OF EVIDENCE

## 1.1 Information in Events

Consider the probability space $(\Omega, \mathcal{A}, P)$, where $\mathcal{A}$ is a $\sigma$-field of sets each of which is a subset of $\Omega$, and the probability of an event $A \in \mathcal{A}$ is $P(A)$. Then we would like a definition for the gain of information about A, which we call $I(A)$, when the event A occurs. $I(A)$ can also be interpreted as the uncertainty concerning the occurrence of A before an experiment is performed. The larger the value of $P(A)$, the less information is gained when A occurs, so we want $I(A)$ to be a decreasing function of $P(A)$. We also desire that the gain in information about the independent events $A_1, A_2, \ldots, A_n$ when they occur, be equal to the sum of the information gain of the individual events (rather than the product, for instance) so that for all n,

$$I\left(\bigcap_{i=1}^{n} A_i\right) = \sum_{i=1}^{n} I(A_i)$$

where

$$P\left(\bigcap_{i=1}^{n} A_i\right) = \prod_{i=1}^{n} P(A_i).$$

Theorem: A function satisfying these two properties is

$$I(A) = -\log P(A).$$

Proof: (i) Suppose $P(A) < P(B)$. Then

$$I(A) = -\log P(A) > -\log P(B) = I(B).$$

(ii) Suppose $A_1, A_2, \ldots$ are independent events.

Then

$$I\left(\bigcap_{i=1}^{n} A_i\right) = -\log P\left(\bigcap_{i=1}^{n} A_i\right) = -\log \prod_{i=1}^{n} P(A_i)$$

$$= -\sum_{i=1}^{n} \log P(A_i) = \sum_{i=1}^{n} I(A_i).$$

Q.E.D.

Any other function satisfying these two properties must be proportional to $-\log P(A)$, the proof of which is a little less simple. Most of the ideas of this section are from Good (1950, pp. 74-75), but they are related to earlier literature.

### 1.1.1 Properties of Information

Some of the properties that $I(A)$ satisfies are as follows:

(i) $I(\Omega) = 0$, that is, no gain in information is obtained when the sure event occurs.

(ii) $I(A) \geq 0$ for all A in $\mathcal{A}$.

(iii) $I(A) = \infty$ if $P(A) = 0$ so that I is unbounded.

## 1.2 Conditional Information in Events

The conditional gain in information $I(A|B)$ is defined as the gain in information when A occurs given that B has occurred, provided that $P(B) > 0$. (This can be expressed timelessly.) Then

$$I(A|B) = -\log P(A|B) = -\log [P(AB)/P(B)]$$
$$= \log P(B) - \log P(AB) = I(AB) - I(B), \qquad (1.2.1)$$

where AB is short for $A \cap B$.

### 1.2.1 Properties of Conditional Information

Some of the properties of $I(A|B)$ are as follows.

(i) If $B \subseteq A$ then $I(A|B) = 0$. Intuitively, this makes sense since the occurrence of B provides all the information about the occurrence of A.

(ii) If $A \subseteq B$ then $I(A|B) = I(A) - I(B)$.

(iii) If A and B are independent then $I(A|B) = I(A)$. That is, B gives us no information concerning A.

(iv) If $P(AB) = 0$ then $I(A|B) = \infty$ provided $P(B) > 0$. In particular, this is true when A and B are mutually exclusive.

References for Section 1.2 are, for example, Good (1950) and (1956).

## 1.3 Mutual Information Between Events

The mutual information between A and B is defined as

$$I(A:B) = I(A) - I(A|B) = I(A) + I(B) - I(AB)$$
$$= I(B) - I(B|A) = I(B:A)$$

from (1.2.1) provided that $P(A) > 0$ and $P(B) > 0$. Thus $I(A:B)$ is symmetric in A and B.

Intuitively, this is a measure of the decrease (if positive) or increase (if negative) in the uncertainty about the occurrence of A (or B) caused by the occurrence of B (or A). This can also be thought of as a measure of the decrease (if positive) or the increase (if negative) in the gain in information when A (or B) occurs, caused by the occurrence of B (or A).

Therefore, $I(A:B)$ can be thought of as a measure of the positive or negative information concerning the occurrence of A (or B) provided by the occurrence of B (or A). $I(A:B)$ can also be expressed as

$$I(A:B) = \log \frac{P(A|B)}{P(A)} = \log \frac{P(AB)}{P(A)P(B)} = \log \frac{P(B|A)}{P(B)} \quad . \tag{1.3.1}$$

It follows that $I(A:B)$ is positive if and only if $P(A|B) > P(A)$ or equivalently $P(B|A) > P(B)$.

### 1.3.1 Properties of Mutual Information

Some of the properties of $I(A:B)$ are as follows.

(i) If A and B are independent then $I(A:B) = 0$ since the occurrence of B gives us no information about the occurrence of A.

(ii) If $B \subseteq A$ then $I(A:B) = I(A)$ since the occurrence of B provides all the information about the occurrence of A.

(iii) If $P(A) > 0$, $P(B) > 0$, and $P(AB) = 0$ then $I(A:B) = -\infty$ which would be the case if A and B were mutually exclusive.

For a further discussion of the ideas in Section 1.3 see Good (1961a).

## 1.4 Weight of Evidence

The expression "weight of evidence" has been used independently by Good (1950), Peirce (1878), and Minsky and Selfridge (1961). Most of the results of Section 1.4 are from Good (1950) and Kullback (1959). Weight of evidence will now be defined.

### 1.4.1 Definitions and Expressions

Let $H_1$ and $H_2$ be two competing hypotheses related to some evidence B. $H_1$ and $H_2$ may be thought of as "events" both in the a priori probability spaces $(\Omega,\mathcal{H},P)$ and $(\Omega,\mathcal{H},P_B)$, where $P_B$ is the conditional probability given B. Then the weight of evidence in favor of $H_1$ as opposed to $H_2$, provided by B, may be defined as

$$W(H_1/H_2:B) = \log \frac{O(H_1/H_2|B)}{O(H_1/H_2)}, \tag{1.4.1.1}$$

where $O(H_1/H_2|B)$ is the odds in favor of $H_1$ as opposed to $H_2$ given B, and $O(H_1/H_2)$ is the odds in favor of $H_1$ as opposed to $H_2$. Weight of evidence may be positive or negative.

Another expression for $W(H_1/H_2:B)$ is, from (1.4.1.1) and (1.3.1),

$$\begin{aligned} W(H_1/H_2:B) &= \log \frac{P(H_1|B)}{P(H_2|B)} - \log \frac{P(H_1)}{P(H_2)} \\ &= \log \frac{P(H_1|B)}{P(H_1)} - \log \frac{P(H_2|B)}{P(H_2)} \\ &= I(H_1:B) - I(H_2:B). \end{aligned} \tag{1.4.1.2}$$

Therefore, the weight of evidence may be interpreted as the difference in information about $H_1$ compared to $H_2$ provided by B.

### 1.4.2 Another Expression for the Weight of Evidence

If W is interpreted as the difference in information about B provided by

$H_1$ compared to $H_2$ then it can be expressed from equation (1.4.2.1) as follows, where B is an event in the probability spaces $(\Omega, \mathcal{B}, P_{H_1})$ and $(\Omega, \mathcal{B}, P_{H_2})$.

$$W(H_1/H_2:B) = \log \frac{P(B|H_1)}{P(B)} - \log \frac{P(B|H_2)}{P(B)} = \log \frac{P(B|H_1)}{P(B|H_2)} \tag{1.4.2.2}$$

which is the log of the likelihood ratio when the hypotheses are simple. Otherwise the non-Bayesian does not necessarily regard the probabilities $P(B|H_1)$ and $P(B|H_2)$ as meaningful. It is historically interesting that the expression "weight of evidence", in its technical sense, anticipated the term "likelihood" by over forty years.

W may also be expressed as

$$W(H_1/H_2:B) = I(B:H_1) - I(B:H_2) = I(B|H_2) - I(B|H_1) \tag{1.4.2.2}$$

because

$$I(B:H_i) = I(B) - I(B|H_i) \, (i = 1,\ 2)$$

from Section 1.3.

### 1.4.3 A Special Case

When $H_2 = H_1^c$ (the complement of $H_1$), then $W(H_1/H_2:B)$ may be abbreviated by $W(H_1:B)$ and the expression in (1.4.1.1) becomes

$$W(H_1:B) = \log \frac{O(H_1|B)}{O(H_1)} = \log \frac{P(B|H_1)}{P(B|H_1^c)},$$

where $O(H_1|B)$ is the odds in favor of $H_1$ given B and $O(H_1)$ is the odds in favor of $H_1$. Good (1950, p. 63, and 1969, p. 25) calls $\frac{O(H_1|B)}{O(H_1)}$ the "Bayes-Jeffreys-Turing factor" in favor of $H_1$ provided by B.

# II. ENTROPY

The entropy I(X) of a random variable or random vector is discussed in this chapter. For the discrete case, it is shown that the entropy may be finite, even when the random variable takes on a denumerable number of values, as shown in a quantum mechanics model in Section 2.1.2. For the continuous case, however, the entropy is always infinite or undefined. Conditional entropy is also discussed.

## 2.1 Entropy of Discrete Random Variables

Let X be a discrete random variable. Then the expected gain in information from a single observation (or the expected uncertainty before an observation is taken) is

$$I(X) = E[I(x)] = -\sum_{i=1}^{\infty} p_i \log p_i ,$$

[Shannon (1948, p. 50)], where E stands for the expected value operation and $p_i = P[X = x_i]$ $(i = 1, 2, \ldots)$. I(X) is called the entropy of X since it is analogous to the average amount of disorder in a system of possible states in thermodynamics. [See, for instance, Slater (1939, p. 33).]

### 2.1.1 Properties of Entropy

Most of these properties are from Shannon (1948, pp. 49-52) except for (i), (vii) and (x).

(i) If the variable takes a denumerable number of values, the entropy may be infinite. [Balakrishnan (1968, p. 193).]

(ii) If the variable takes a finite number of values, then the entropy is a maximum when all the $p_i$'s are equal.

(iiia) I(X) is a continuous function of the $p_i$'s.

(iiib) If all the $p_i$'s are equal to $\frac{1}{n}$ then

$$I(X) = -\sum_{i=1}^{n} \frac{1}{n} \log \frac{1}{n} = \log n$$

and this is a monotone increasing function of n.

(iv) I(X) is the variable in Boltzmann's famous theorem in statistical mechanics. See, for instance, Condon and Odishaw (1958, page 5-16).

(v) I(X) = 0 if and only if one of the $p_i$'s is 1.

(vi) $I(X) \geq 0$

(vii) $I(f(X)) \leq I(X)$ for any transformation f(X), with equality if f(X) is one-to-one. [Dobrushin (1963, p. 361).]

(viii) Any change toward equalizing $p_i$ increases I(X). In fact, if

$$p_i' = \sum_{j=1}^{n} a_{ij} p_j$$

where

$$\sum_{i=1}^{n} a_{ij} = \sum_{j=1}^{n} a_{ij} = 1 ,$$

then the use of the $p_i'$'s instead of the $p_i$'s will not decrease the value of I(X).

(ix) If I(X,Y) is defined as

$$E[I(x_i,y_j)] = - \sum_{i=1}^{n} \sum_{j=1}^{m} p_{ij} \log p_{ij}$$

where

$$p_{ij} = P[X = x_i, Y = y_j] ,$$

then

$$I(X,Y) \leq I(X) + I(Y) ,$$

with equality if and only if X and Y are independent.

(x) If $\mathbf{X} = (X_1, X_2, \ldots)$ then

$$I(\mathbf{X}) = \lim_{n\to\infty} I((X_1, \ldots, X_n))$$

from Pinsker (1964, p. 11) where $X_1, X_2, \ldots$ are random variables.

### 2.1.2 Quantum Mechanics Model

An example of an infinite-state case [Balakrishnan (1968, p. 193)] is the quantum-mechanical model where the variable is the number of photons in a given mode. Let $p_n$ be the probability of exactly n photons where the average number of photons is $\sum_{n=0}^{\infty} n\, p_n$ . If $\nu$ is the mode frequency, and h is Planck's constant, and the average energy is E, then $E/h\nu = \sum_{n=0}^{\infty} np_n$ and we may consider the problem of determining the photon distribution corresponding to the maximum entropy $-\sum_{n=0}^{\infty} p_n \log p_n$ for a fixed energy E, which leads to

$$p = e^{-n\alpha}/(1 - e^{-\alpha}) = \frac{h\nu}{E}\left(1 + \frac{h\nu}{E}\right)^{-n-1}$$

(where $\alpha$ is a positive constant) and the corresponding entropy is always finite. [See Takahasi (1965).]

## 2.2 Conditional Entropy for Discrete Random Variables

Suppose X and Y are discrete random variables. Then define the entropy of X given Y, where $p_j(i) = P[X = x_i\ Y = y_j]$, as

$$I[X|Y] = E_{XY}[I(x|y)] = \sum_{i=1}^{n} \sum_{j=1}^{m} p_{ij} \log p_j(i) \;.$$

This will be used in the definition of expected mutual information in Section 4.1.1.

### 2.2.1 Some Properties of Conditional Entropy

(i) $I(X,Y) = I(X) + I(Y|X) = I(Y) + I(X|Y)$

(ii) $I(Y) \geq I(Y|X)$

The results of 2.2 are from Shannon (1948, p. 52).

## 2.3 Entropy of Random Variables with Continuous Probability Density Functions

Most of the results of this section are from Reza (1961, p. 271).

Let X be a random variable with a continuous density function and let [a,b] be any closed interval over which X is defined. Dissect the interval into n parts $\Delta x_k (k = 1,\ldots,n)$ so that $b - a = \sum_{k=1}^{n} \Delta x_k$ and find $p_k$'s so that

$$P[a \le X \le b] = \sum_{k=1}^{n} p_k \, \Delta x_k$$

where $p_k \, \Delta x_k$ is the probability of X falling in the kth subinterval. Then

$$I(X) = E\, I(x) = \lim_{\substack{b\to\infty \\ a\to-\infty}} \; \lim_{\substack{n\to\infty \\ \text{Max}\, \Delta x_k \to 0}} \left[ - \sum_{k=1}^{n} (p_k \, \Delta x_k) \cdot \log (p_k \, \Delta x_k) \right]$$

$$= \lim_{\substack{b\to\infty \\ a\to-\infty}} \left[ - \int_a^b f(x) \log f(x) dx - \lim_{\substack{n\to\infty \\ \text{Max}\, \Delta x_k \to 0}} \sum_{k=1}^{n} (p_k \, \Delta x_k) \log \Delta x_k \right] ,$$

where f(x) is the continuous probability density function.

For any $M > 0$, there exists an m such that

$$\log \text{Max}\, \Delta x_k < \frac{-M}{P[a \le X \le b]}$$

where $b - a = \sum_{k=1}^{m} \Delta x_k$, and hence,

$$\sum_{k=1}^{m} p_k \, \Delta x_k \log \Delta x_k < \sum_{k=1}^{m} p_k \, \Delta x_k (-M/P[a \le X \le b]) = -M \, .$$

Therefore,

$$-I(X) < + \int_{-\infty}^{\infty} f(x) \log f(x) dx - M$$

so that

$$I(X) > - \int_{-\infty}^{\infty} f(x) \log f(x) dx + M$$

for any $M > 0$ and so I(X) is either undefined or unbounded although $h(X) = -\int_{-\infty}^{\infty} f(x) \log f(x) dx$, which Kolmogorov (1956) calls the differential entropy, may be finite. Then $I(X) = \infty$. Now h(X) has no direct interpretation and may be positive or negative or equal to $-\infty$ or $+\infty$ and is not invariant with respect to one-to-one coordinate transformations $Y = t(X)$. [Good (1968b and 1968c) introduces a definition for an invariantized entropy.] For such a transformation

$$h(Y) = h(X) - E[\log J( \frac{X}{Y} )]$$

where $J(\frac{X}{Y})$ is the Jacobian of X with respect to Y. See Shannon (1948, p. 90). h(X) has no analogue for infinite-dimensional distributions as pointed out by Kolmogorov (1956).

## 2.4 Conditional Entropy of Random Variables with Continuous Density Functions

Let X and Y be random variables with continuous density functions. We naturally try to define the conditional entropy $I(X|Y)$ as the expectation with respect to Y of $I(X|y)$, that is, of the entropy of X for a fixed value of Y. But it can be seen, along the lines of Section 2.3, that the result is either meaningless or infinite.

# III. SINGULARITY BETWEEN TWO PROBABILITY MEASURES

Before defining expected mutual information we must first investigate the relationship of singularity between two probability measures. We shall consider univariate, bivariate, and multivariate distributions.

Let $P_0$ and $P_1$ represent two probability measures over the same measurable space $(\Omega, \mathcal{A})$ where $\mathcal{A}$ is a complete $\sigma$-field of sets in $\Omega$ (where "complete" means that the $\sigma$-field contains all subsets of sets of measure zero with respect to $P_0$ and $P_1$). Then from the Lebesgue Decomposition Theorem and the Radon-Nikodym Theorem it follows that there exists a set B in $\mathcal{A}$ where $P_1(B) = 0$ such that for any set A in $\mathcal{A}$

$$P_0(A) = \int_A f dP_1 + P_0(AB) \qquad (3.0.1)$$

where f is a non-negative function defined on $\Omega$, and is also measurable, which means that for any Borel set C, $f^{-1}(C)$ is in $\mathcal{A}$.

The references on these topics and the definitions of singularity in the following three sections, 3.1 - 3.3, are taken from Doob (1953, p. 611), Grenander (1950, p. 210), Yaglom (1962, pp. 327, 328), and Loève (1963, pp. 300-302).

## 3.1 Definitions

A few relationships between $P_0$ and $P_1$ will now be examined.

### 3.1.1 Regular and Parallel Cases

If $P_0(B) = 0$ whenever $P_1(B) = 0$, we have what Yaglom calls the regular case. $P_0$ is then said to be absolutely continuous with respect to $P_1$, the last term in (3.0.1) vanishes, and f is called the Radon-Nikodym derivative of $P_0$ with respect to $P_1$. It is then natural to say that $f = \frac{dP_0}{dP_1}$ almost everywhere with respect to $P_1$, and we have

$$P_0(A) = \int_A f dP_1 = \int_A \frac{dP_0}{dP_1} dP_1 = \int_A dP_0 .$$

If $P_0(A) = \int_A f dP_1 = 0$ then $P_1(A) = \int_A dP_1 = 0$ unless f can equal zero on sets where $P_1$ is not zero. If f is not zero on sets where $P_1$ is not zero (i.e. $f \neq 0$ a.e. with respect to $P_1$) then $P_1$ is also absolutely continuous with respect to $P_0$. In this case $P_0$ and $P_1$ are often said to be "equivalent", but we regard this as an unfortunate term and we replace it by parallel and we write $P_0 \parallel P_1$.

### 3.1.2 Intermediate Singular Case

If $0 < P_0(B) < 1$, but $P_1(B) = 0$ we have what Yaglom calls the intermediate singular case (of $P_0$ with respect to $P_1$). $P_0$ is not absolutely continuous with respect to $P_1$, but $P_1$ may or may not be absolutely continuous with respect to $P_0$.

### 3.1.3 Extreme Singular or Perpendicular Case

If $P_0(B) = 1$ and $P_1(B) = 0$ for some B in $\mathcal{A}$, we have what Yaglom calls the extreme singular case and $P_0$ and $P_1$ are said to be perpendicular (written $P_0 \perp P_1$). Then $P_0(B^c) = 0$ and $P_1(B^c) = 1$.

## 3.2 Examples

### 3.2.1 Discrete Uniform Probabilities

Let $P_0$ and $P_1$ be two discrete probability measures with mass functions $p_0(x)$ and $p_1(x)$ defined as follows. Let $p_0(x) = \frac{1}{3}$ if x = 1, 2, or 3 and $p_1(x) = \frac{1}{2}$ if x = 1 or 2. $P_0$ is not absolutely continuous with respect to $P_1$ since $p_0(3) = \frac{1}{3}$ and $p_1(3) = 0$, but $P_1$ is absolutely continuous with respect to $P_0$, that is, $P_1 = 0$ whenever $P_0 = 0$. Consider the equation (3.0.1). Let $B = \{3\}$, the set containing only the element 3, and let $f = \frac{p_0(x)}{p_1(x)}$ when $x \in \{1,2\}$ and let f = 0 elsewhere. Then (3.0.1) becomes

$$P_0(A) = \sum_{x \text{ in } A} f\, p_1(x) + P_0(AB) = \sum_{x \text{ in } A\cap\{1,2\}} \frac{p_0(x)}{p_1(x)} p_1(x) + P_0[A \cap \{3\}]$$

$$= P_0[A \cap \{1,2\}] + P_0[A \cap \{3\}] ,$$

for any measurable set A, so that (3.0.1) is confirmed. This is an intermediate singular case of $P_0$ with respect to $P_1$ and a nonsingular case of $P_1$ with respect to $P_0$. To use (3.0.1) to find $P_1(A)$, let $f = \frac{p_1(x)}{p_0(x)}$ when x is in {1,2} and let f = 0 elsewhere. Then (3.0.1) becomes

$$P_1(A) = \sum_{x \text{ in } A} \frac{p_1(x)}{p_0(x)} p_0(x) = \sum_{x \text{ in } A} p_1(x) ,$$

which checks.

### 3.2.2 The Univariate Distribution Function

Let $\mu$ be the Lebesgue measure on the real line R and let P be a probability measure on $(R,\mathcal{B})$ where $\mathcal{B}$ is the $\sigma$-field of Borel sets. Of course

$$\mu(a,x) = \int_{(a,x)} d\mu = x - a = \int_a^x dt,$$

and (3.0.1) reduces to

$$P[(-\infty,x)] = \int_{-\infty}^x f(t)dt + P[(-\infty,x) \cap (D \cup C)] ,$$

where D is the set of all points $t_i$ $(i = 1, 2, \ldots)$ for which $P(\{t_i\}) > 0$, and C is a set of Lebesgue measure zero, containing no point $t_i$, yet possibly having positive probability measure. Thus

$$P[(-\infty,x)] = \int_{-\infty}^x f(t)dt + P[(-\infty,x) \cap C] + \sum_{t_i \le x} P(\{t_i\}) . \quad (3.2.2.1)$$

Let

$$\gamma_1 = \int_{-\infty}^{\infty} f(t)dt, \ \gamma_2 = P[(-\infty,\infty) \cap C], \ \gamma_3 = \sum_{t_i \le \infty} P(\{t_i\})$$

then $\gamma_1 + \gamma_2 + \gamma_3 = 1$. Let ac denote "absolutely continuous", sc "singularly continuous", and d "discrete". Define

$$F(x) = P[(-\infty,x)] ,$$

$$F_{ac}(x) = (1/\gamma_1) \int_{-\infty}^x f(t)dt ,$$

$$F_{sc}(x) = (1/\gamma_2) P[(-\infty,x) \cap C] ,$$

and

$$F_d(x) = (1/\gamma_3) \sum_{t_i \le x} P(\{t_i\}) .$$

Then $F(x)$, $F_{ac}(x)$, $F_{sc}(x)$, and $F_d(x)$ are distribution functions and the Equation (3.2.2.1) can be written

$$F(x) = \gamma_1 F_{ac}(x) + \gamma_2 F_{sc}(x) + \gamma_3 F_d(x) . \qquad (3.2.2.2)$$

Cramér (1937, p. 17) mentioned that such a combination is always possible, with references to E. W. Hobson. The following cases will now be investigated.

(i) $P(D \cup C) = 0$, when P is absolutely continuous and $F(x) = F_{ac}(x)$.

(ii) $0 < P(D \cup C) < 1$, the intermediate singular case.

(iii) $P(D \cup C) = 1$, the extreme singular case where

$$F(x) = \gamma_2 F_{sc}(x) + \gamma_3 F_d(x) \quad (\gamma_2 + \gamma_3 = 1) .$$

(iiia) $P(D) = 1$, the discrete case, where $F(x) = F_d(x)$.

(iiib) $P(C) = 1$, then all the probability is contained in a set of Lebesgue measure zero, there is no single point having positive probabilities, and $F(x) = F_{sc}(x)$. For example, $F_{sc}(x)$ might be the Cantor function which is constant on intervals, but has a different value on each interval. On the other hand $F_{sc}(x)$ might not be constant on any interval, even though the density function $f(x)$ is zero almost everywhere. These types of functions are discussed, for example, by Munroe (1953, pp. 193, 287).

### 3.2.3 Singular Continuous Bivariate Distribution Function

Let X be a random variable with a continuous density function and let $Y = X$. The joint probability $P_{XY}$ satisfies $P_{XY}(B) = 1$, where $B = \{(x,y) | x = y\}$. Then, for any measurable set A in the XY plane, (3.0.1) becomes $P_{XY}(A) = P_{XY}(AB)$, where B has zero Lebesgue measure in the XY plane, the joint distribution function is

$$F_{XY}(a,b) = F_X(\text{Min }\{a,b\}) = F_Y(\text{Min }\{a,b\}) .$$

$F_{XY}(x,y)$ is a valid distribution function being a continuous monotone non-decreasing function of x and y with $F_{XY}(-\infty, -\infty) = F_X(-\infty) = 0$ and $F_{XY}(\infty,\infty) = F_X(\infty) = 1$.

It is easily seen that, when (a,b) is any point not on the line X = Y,

$$f_{XY}(a,b) = \frac{\partial^2}{\partial x \partial y} F_{XY}(x,y) \Big|_{x = a,\ y = b} = 0 ,$$

but $f_{XY}(a,a)$ is undefined when $f_X(a) > 0$. Even $\frac{\partial}{\partial x} F_{XY}(x,a) \Big|_{x = a}$ is undefined since $[F_{XY}(a,a) - F_{XY}(a,a)]/(x-a)$ tends to two distinct limits as $x \to a+0$, and $x \to a-0$, these limits being 0 and $f_X(a)$ respectively.

## 3.3 Singularity between the Joint Probability and the Product Probability Measures

Let $(\Omega,\mathcal{A}_1,P_1)$ and $(\Omega_2,\mathcal{A}_2,P_2)$ be two probability spaces. The product probability measure $P_1 \times P_2$ is defined on the measurable space $(\Omega_1 \times \Omega_2, \mathcal{A}_1 \times \mathcal{A}_2)$ by the equation

$$P_1 \times P_2(A_1 \times A_2) = P_1(A_1)P_2(A_2)$$

whenever $A_1 \varepsilon \mathcal{A}_1$ and $A_2 \varepsilon \mathcal{A}_2$.

### 3.3.1 Bivariate Example

Let X and Y be the two (identical) random variables defined in Section 3.2.3. Define the product probability $P_X \times P_Y$ on the measurable space $(R_X \times R_Y, \mathcal{B}_X \times \mathcal{B}_Y)$, where $(R_X, \mathcal{B}_X)$ and $(R_Y, \mathcal{B}_Y)$ are the Borel measurable spaces of X and Y. If we think of dx and dy as sets as well as differentials, we have

$$P_X \times P_Y(dx \times dy) = P_X(dx)P_Y(dy) = f_X(x)dx\ f_Y(y)dy,$$

so that $P_X \times P_Y$ is absolutely continuous with respect to two-dimensional

Lebesgue measure in the XY plane. Clearly

$$P_X \times P_Y(C) = \int_C \int f_X(x) f_Y(y)\,dxdy$$

for any C in $\mathcal{B}_X \times \mathcal{B}_Y$ from (3.0.1). The distribution function of $P_X \times P_Y$ is $F_X \times F_Y(x,y) = F_X(x)F_Y(y)$ and the density function is $f_X \times f_Y(x,y) = f_X(x)f_Y(y)$. If $C_1 \varepsilon \mathcal{B}_X$ and $C_2 \varepsilon \mathcal{B}_Y$, then

$$P_X \times P_Y(C_1 \times C_2) = \int_{C_1} f_X(x)\,dx \int_{C_2} f_Y(y)\,dy.$$

Consider the set $B = \{(x,y) \,|x = y\}$ defined in Section 3.2.3. Then $P_{XY}(B) = 1$ and $P_X \times P_Y(B) = 0$ so that $P_{XY} \perp P_X \times P_Y$, as defined in Section 3.1.3. Therefore, $P_{XY}$ is singular with respect to $P_X \times P_Y$.

3.3.2 A Multidimensional Case

(i) Let $\mathbf{X}' = (X_1, \ldots, X_k)$ and $\mathbf{Y}' = (Y_1, \ldots, Y_\ell)$ represent multivariate random vectors. Consider the hypersurface $B = \{(\mathbf{x},\mathbf{y}) \,|x_i = \Theta(\mathbf{y})\}$ for some component $X_i = x_i$ and some function $\Theta$ mapping the $\ell$-dimensional into 1-dimensional space. Suppose $P_{\mathbf{XY}}(A) > 0$ for some $A \subseteq B$. Let $\mathbf{X}$ and $\mathbf{Y}$ have continuous density functions on the perpendicular projections $A_{\mathbf{X}}$ and $A_{\mathbf{Y}}$ onto the $\mathbf{X}$ and $\mathbf{Y}$ spaces, respectively. Then

$$P_{\mathbf{X}} \times P_{\mathbf{Y}}(C) = \iint_C f_{\mathbf{X}}(x) f_{\mathbf{Y}}(y)\,dxdy\ ,$$

where $C \varepsilon A_{\mathbf{X}} \times A_{\mathbf{Y}}$ and $P_{\mathbf{X}} \times P_{\mathbf{Y}}(A) = 0$. Therefore, $P_{\mathbf{XY}}$ is not absolutely continuous with respect to $P_{\mathbf{X}} \times P_{\mathbf{Y}}$ since $P_{\mathbf{X}} \times P_{\mathbf{Y}}(A) = 0$ does not imply $P_{\mathbf{XY}}(A) = 0$. In other words $P_{\mathbf{XY}}$ is singular with respect to $P_{\mathbf{X}} \times P_{\mathbf{Y}}$. For any measurable set E, the equation in (3.0.1) is then

$$P_{\mathbf{XY}}(E) = \iint_{ED^c} f\, d(P_{\mathbf{X}} \times P_{\mathbf{Y}}) + P_{\mathbf{XY}}(ED)\ ,$$

where D is a hypersurface in the $(k + \ell)$-dimensional space, and

$$P_{\mathbf{X}} \times P_{\mathbf{Y}}(D) = 0, \quad g = \frac{f(\mathbf{x},\mathbf{y})}{f(\mathbf{x})f(\mathbf{y})}\ ,$$

and

$$d(P_{\mathbf{X}} \times P_{\mathbf{Y}}) = f(\mathbf{x})f(\mathbf{y})dxdy .$$

It follows that

$$P_{\mathbf{XY}}(E) = \iint_{ED^c} f(\mathbf{x},\mathbf{y})dx\,dy + P_{\mathbf{XY}}(ED) .$$

An example where $P_X \times P_Y$ is not absolutely continuous with respect to $P_{XY}$ even when

$$P_{XY}(E) = \iint_E f(x,y)dxdy$$

for all measurable sets E, is as follows.

(ii) Define $F_{XY}(x,y)$ as

$$f_{XY}(x,y) = \begin{cases} 0 & \text{if } 0 \le x \le 1,\ 0 \le y \le 1, \\ \frac{1}{3} & \text{if } 0 \le x \le 2,\ 1 \le y \le 2, \\ \frac{1}{3} & \text{if } 1 \le x \le 2,\ 0 \le y \le 1. \end{cases}$$

Let $A = \{(x,y)\,|\,0 \le x \le 1,\ 0 \le y \le 1\}$. Then $P_{XY}(A) = 0$, but $P_X \times P_Y(A) = 1/9 \neq 0$. For any measurable set E, where $P_{XY}(A) = 0$, Equation (3.0.1) now becomes

$$\begin{aligned} P_X \times P_Y(E) &= \iint_{EA^c} f\, dP_{XY} + P_X \times P_Y(EA) \\ &= \iint_{EA^c} f_X(x)f_Y(y)dx\,dy + \iint_{EA} f_X(x)f_Y(y)dx\,dy \end{aligned}$$

because $P_X \times P_Y$ is absolutely continuous with respect to two-dimensional Lebesgue measure.

(iii) If $P_{XY}$ and $P_X \times P_Y$ are both absolutely continuous with respect to two-dimensional Lebesgue measure, and $f(x,y) > 0$ if and only if $f(x)f(y) > 0$, as with nonsingular normal variables, then $P_{XY}$ is parallel with $P_X \times P_Y$ (as defined in Section 3.1.1).

### 3.3.3 The Multidimensional Normal Distributions

Let $\mathbf{X}' = (X_1, \ldots, X_k)$ and $\mathbf{Y}' = (Y_1, \ldots, Y_\ell)$ be multidimensional normal random vectors. Then the joint characteristic function of $\mathbf{X}$ and $\mathbf{Y}$, defined as

$$\phi(\mathbf{t},\mathbf{s}) = E[\exp i\,(\mathbf{t}'\mathbf{x} + \mathbf{s}'\mathbf{y})]\ ,$$

where **t** and **x** are k-dimensional column vectors and **s** and **y** are $\ell$-dimensional column vectors, is equal to

$$\phi(\mathbf{t},\mathbf{s}) = \exp[i(\mu_X' \mathbf{t} + \mu_Y' \mathbf{s}) - \frac{1}{2}(\mathbf{t},\mathbf{s})'\mathbf{A}(\mathbf{t},\mathbf{s})]$$

where $\mu_X$ and $\mu_Y$ are the mean vectors of **X** and **Y**, and **A** is the covariance matrix. If **A** has full rank $k + \ell$ then $P_{\mathbf{XY}}$ is absolutely continuous over the entire $k + \ell$ space, so that for any measurable set S,

$$P_{\mathbf{XY}}(S) = \int \underset{S}{\cdots} \int f(\mathbf{x},\mathbf{y})\, d\mathbf{x}\, d\mathbf{y}.$$

Therefore, $P_{\mathbf{XY}}$ cannot have positive probability on any hyperplane. If **A** has rank r, where $r < k + \ell$ then $P_{\mathbf{XY}}$ is equal to 1 on some r-dimensional submanifold of the form

$$\sum_{i=1}^{k} (a_i X_i + c_i) + \sum_{i=1}^{\ell} (b_i y_i + d_i) = 0,$$

where $a_i$, $c_i$, $b_i$, and $d_i$ are real constants. [Cramér (1946, p. 312).]

It therefore follows that in this case either $P_{\mathbf{XY}} \perp P_{\mathbf{X}} \times P_{\mathbf{Y}}$ or $P_{\mathbf{XY}} \parallel P_{\mathbf{X}} \times P_{\mathbf{Y}}$ (as defined in Section 3.1.1) and the intermediate singular case never occurs.

# IV. EXPECTED MUTUAL INFORMATION

The expected mutual information between two random vectors is discussed in this chapter. This is infinite if the joint probability is not absolutely continuous with respect to the product probability. The special case where the random vectors are normally distributed is also discussed.

## 4.1 Discrete or Absolutely Continuous Random Variables X and Y

### 4.1.1 Definitions

The expected mutual information may be defined as

$$I(X:Y) = I(X) - I(X|Y) = I(X) + I(Y) - I(X,Y) = I(Y) - I(Y|X) = I(Y:X)$$

in the discrete case, from (2.2.1) (Shannon, 1948).

This reduces to

$$I(X:Y) = h(X) - h(X|Y) = h(X) + h(Y) - h(X,Y) = h(Y) - h(Y|X) = I(Y:X)$$

in the continuous case, from Sections 2.3 and 2.4, since the latter part of the expressions for $I(X)$ and $I(X|Y)$ in Sections 2.3 and 2.4 will cancel. (Reza, 1961, p. 276.)

Therefore $I(X:Y)$ is symmetric in X and Y. $I(X:Y)$ may be interpreted as the decrease in entropy, or the decrease in the expected gain in information, or the decrease in the expected uncertainty of X (or Y) when Y (or X) is observed. Thus $I(X:Y)$ can be thought of as the expected information about X (or Y) provided by Y (or X).

### 4.1.2 Various Expressions for I(X:Y)

If X and Y are discrete then, from the above,

$$I(X:Y) = -E[\log p_i] + E[\log p_j(i)] = E[\log ( \frac{p_{ij}}{p_i\ p_j} )]$$

$$= E[\log \frac{p_i(j)}{p_j} ] = I(Y:X) \qquad (4.1.2.1)$$

where $p_i = P[X = x_i]$, $p_j = P[Y = y_j]$, $p_{ij} = P[X = x_i,\ Y = y_j]$,

$p_j(i) = P[X = x_i | Y = y_j]$, etc.

If X and Y have continuous probability density functions then similarly

$$I(X:Y) = -E_X[\log f_X(x)] + E_{XY}[\log f_{x|y}(x|y)]$$

$$= E_{XY}[\log \frac{f_{XY}(x,y)}{f_X(x) f_Y(y)}] = E_{XY}[\log \frac{f_{Y|X}(y|x)}{f_Y(y)}] = I(Y:X) \ . \qquad (4.1.2.2)$$

4.1.3 I(S:R) between the Received and Original Signals

Let R = S + N where S and N are absolutely continuous independent random variables. Then

$$I(S:R) = I(S:S + N) = h(S + N) - h(S + N|S)$$

$$= h(S + N) - \int_{-\infty}^{\infty} f_S(s) ds \int_{-\infty}^{\infty} f_{R|S}(s + n|s) \log f_{R|S}(s + n|s) dn \ .$$

Since the distribution functions satisfy

$$F_{R|S}(s + n|s) = P[S + N \le s + n | S = s] = P[N \le n] = F_N(n),$$

it follows that $f_{R|S}(s + n|s) = f_N(n)$ and therefore,

$$I(S:R) = h(S + N) - \int_{-\infty}^{\infty} f_S(s) ds \int_{-\infty}^{\infty} f_N(n) \log f_N(n) dn$$

$$= h(S + N) - h(N) = h(R) - h(N) \ . \qquad (4.1.3.1)$$

See, for instance, Shannon (1948) or Reza (1961, p. 285).

4.2 Multidimensional Random Vectors where the Components may be Discrete, Continuous, or a Mixture of Both

This section is an extension of a method which is mentioned in Gelfand and Yaglom (1959, pp. 202-205). It may be ignored without loss of continuity since the general definition of expected mutual information given in Section 4.3 includes this case also. Theorem 4.2 is a special case where $P_{XY}$ is not absolutely continuous with respect to $P_X \times P_Y$, as pointed out in Section 3.2.2, and therefore, $I(X:Y) = \infty$ as mentioned in the first part of Section 4.5. This section has been included to illustrate a method of partitioning the range of **X** and **Y** by transforming the vectors into integer scalars.

Suppose $\mathbf{X}' = (X_1, \ldots, X_k)$ and $\mathbf{Y}' = (Y_1, \ldots, Y_\ell)$ where the components may be discrete, continuous, or a mixture of both. Partition the ranges of **X** and **Y** into a finite number of non-overlapping (k-dimensional or less) and ($\ell$-dimensional or less) blocks $\Delta_1, \ldots, \Delta_n$ and $\Delta_1', \ldots, \Delta_m'$, where the edges of the blocks may be finite or infinite, open or closed, or semiclosed. Consider now the discrete scalar random variables $X(\Delta_1, \ldots, \Delta_n)$ and $Y(\Delta_1', \ldots, \Delta_m')$ whose values are equal to the subscripts of those blocks $\Delta_i$ $(i = 1,2,\ldots n)$ and $\Delta_j$ $(j = 1,2,\ldots m)$ to which the values of **X** and **Y** belong. [(E.g. if $\mathbf{X} = (1.1, 3.5, \ldots, 6.1) \varepsilon \Delta_3$, then $X(\Delta_1,\ldots, \Delta_n) = 3$).]

It is shown by Gelfand and Yaglom (1959) that

$$I[X(\Delta_1^*, \ldots, \Delta_r^*):Y(\Delta_1', \ldots, \Delta_m')] \geq I[X(\Delta_1, \ldots, \Delta_n):Y(\Delta_1', \ldots, \Delta_m')]$$

whenever $\Delta_1^*, \ldots, \Delta_r^*$ (where $n \leq r$) is a refinement of $\Delta_1, \ldots, \Delta_n$. That is, each interval $\Delta_i^* \varepsilon \Delta_j$ for some $j = 1, \ldots, n$, and the $\Delta^*$'s span the k-dimensional space. Therefore measuring **X** or **Y** more precisely can only increase the mutual information.

$I(\mathbf{X}:\mathbf{Y})$ is defined as follows:

$$I(\mathbf{X}:\mathbf{Y}) = \sup_{\substack{\Delta_i, \Delta_j' \\ m,n}} I[X(\Delta_1, \ldots, \Delta_n):Y(\Delta_1', \ldots, \Delta_m')]$$

$$= \lim_{\substack{\max \Delta_i \to 0 \\ \max \Delta_j' \to 0 \\ n,m\to\infty}} I[X(\Delta_1, \ldots, \Delta_n):Y(\Delta_1', \ldots, \Delta_m')] \; .$$

Let $X_n = X(\Delta_1, \ldots, \Delta_n)$ and $Y_m = Y(\Delta_1', \ldots, \Delta_m')$. Then

$$I(X_n:Y_m) = \sum_{i=1}^{n} \sum_{j=1}^{m} p_{X_nY_m}(i,j) \log[p_{X_nY_m}(i,j)/p_{X_n}(i)p_{Y_m}(j)],$$

where

$$\sum_{i=1}^{n} \sum_{j=1}^{m} p_{X_nY_m}(i,j) = 1, \quad p_{X_nY_m}(i,j) = P[X_n = i,\ Y_m = j],$$

$$p_{X_n}(i) = P[X_n = i], \quad \text{and} \quad p_{Y_m}(j) = P[Y_m = j] .$$

The following theorem shows that expected mutual information is infinite when $P_{\mathbf{XY}}$ is not absolutely continuous with respect to $P_{\mathbf{X}} \times P_{\mathbf{Y}}$.

Theorem 4.2 Let $\mathbf{X}$ and $\mathbf{Y}$ be defined as in Section 3.3.2 (i). Then $I(\mathbf{X}:\mathbf{Y}) = \infty$.

Proof: Partition the subset A into s equiprobable parts $A_j$ $(j = 1, \ldots, s)$ which will always be possible because $F_{\mathbf{XY}}$ is continuous on this subspace. Select $\Delta_1, \ldots, \Delta_s$ and $\Delta_1', \ldots, \Delta_s'$ where $s \leq n$ and $s \leq m$ so that the projection of $A_j$ on the k-dimensional space $\mathbf{X}$ is $\Delta_j$ and the projection of $A_j$ on the $\ell$-dimensional space $\mathbf{Y}$ is $\Delta_j'$ $(j = 1, \ldots, s)$. It then follows that $P_{\mathbf{XY}}(A_j) = P_{\mathbf{X}}(\Delta_j) = P_{\mathbf{Y}}(\Delta_j') = p/s$ $(j = 1, \ldots, s)$ where $P_{\mathbf{XY}}(A) = p$. Then

$$I(\mathbf{X}:\mathbf{Y}) \geq \sum_{j=1}^{s} P_{\mathbf{XY}}(A_j) \log \frac{P_{\mathbf{XY}}(A_j)}{P_{\mathbf{X}}(\Delta_j)P_{\mathbf{Y}}(\Delta_j')} = \sum_{j=1}^{s} (p/s) \log \frac{(p/s)}{(p/s)^2}$$

$$= -p \log (p/s) \rightarrow \infty \text{ as } s \rightarrow \infty,$$

so that $I(\mathbf{X}:\mathbf{Y}) = \infty$.

Corollary 4.2 If $\mathbf{X} = (X_1, \ldots, X_k)'$ has a continuous distribution function $F_{\mathbf{X}}$ over some k-dimensional block $\mathbf{A}$ in the k-dimensional $\mathbf{X}$ space and $P_{\mathbf{X}}(A) > 0$, then $I(\mathbf{X}:\mathbf{X}) = \infty$ [Gelfand and Yaglom (1959, p. 205)].

Proof: This follows immediately from the above theorem since $X_i = X_i$ $(i = 1, \ldots, k)$.

Some authors define the entropy of $\mathbf{X}$ to be $I(\mathbf{X}) = I(\mathbf{X}:\mathbf{X})$. See Section 4.4 (x) and (xi).

## 4.3 General Definition for Expected Mutual Information

Let $(\Omega_T, A_T, P_{\mathbf{X}_T})$ and $(\Omega_S, A_S, P_{\mathbf{Y}_S})$ be the probability spaces for the random processes $\mathbf{X}_T = \{X(t) | t \text{ in } T\}$ and $\mathbf{Y}_S = \{Y(s) | s \text{ in } S\}$ where T and S are any ordered sets. T and S may be discrete sets, in which case $\mathbf{X}_T$ and $\mathbf{Y}_S$ are vectors. Throughout most of this paper T and S will represent the time intervals $[0,T]$ and $[0,S]$.

Let $P_{\mathbf{X}_T\mathbf{Y}_S}$ and $P_{\mathbf{X}_T} \times P_{\mathbf{Y}_T}$ represent the joint and product probability measures on the measurable spaces $(\Omega_T \times \Omega_S, A_T \times A_S)$. That is, $\Omega_T \times \Omega_S = \bigcup_{i=1}^{\infty} E_i$, where the $E_i$ are mutually disjoint.

The expected mutual information may then be defined by

$$I(\mathbf{X}_T:\mathbf{Y}_S) = \sup \sum_{i=1}^{\infty} P_{\mathbf{X}_T\mathbf{Y}_S}(E_i) \log \left\{ \frac{P_{\mathbf{X}_T\mathbf{Y}_S}(E_i)}{P_{\mathbf{X}_T} \times P_{\mathbf{Y}_S}(E_i)} \right\} \tag{4.3.1}$$

where the sup is the least upper bound over all partitions of $\Omega_T \times \Omega_S$. [See also Section 4.5.] This definition is from Pinsker (1964, p. 19). If $P_{\mathbf{X}_T\mathbf{Y}_S}(E_i) = 0$, then set the term in the summand equal to 0.

## 4.4 Properties of Expected Mutual Information

In this section the reference "Gelfand and Yaglom (1959)" will be abbreviated by G. & Y. For the following properties $\mathbf{X}$ and $\mathbf{Y}$ may be vectors.

(i) $I(\mathbf{X}:\mathbf{Y}) \geq 0$. If $I(\mathbf{X}:\mathbf{X}) = 0$ then $\mathbf{X}$ and $\mathbf{Y}$ are independent. [G. & Y., pp. 209, 214.]

(ii) $I(\mathbf{X}:\mathbf{Y})$ is invariant under one-one change in coordinates. [G. & Y., pp. 209, 214.]

(iii) $I(\mathbf{X}:\mathbf{Y})$ is well defined for infinite-dimensional distributions. [Pinsker (1964, p. 11).]

(iv)

$$I(X:Y) \geq I(f(X):Y)$$

with equality if f(X) is one-one. [G. & Y., pp. 209, 214] and [Dobrushin (1963, p. 361).]

(v)

$$I[(X_1,X_2):(Y_1,Y_2)] = I(X_1:Y_1) + I(X_2:Y_2)$$

if $(X_1,Y_1)$ and $(X_2,Y_2)$ are independent. [ Pinsker (1964, p. 11).]

(vi)

$$I[(X,Y):Z] \geq I(X:Z)$$

so that I(X:Y) can only increase with more components in X or Y. [G. & Y., p. 206.]

(vii)

$$I[(X,Y):Z] = I(X:Z)$$

if and only if X, Y, Z form a Markov sequence. That is if

$$P[Z \leq z | X = x, Y = y] = P[Z \leq z | Y = y]$$

[Kolmogorov (1956, p. 103)].

(viii) I(X:Y) = 0 and I(Z:Y) = 0 together do not imply that I[(X,Z):Y] = 0 [Kolmogorov (1956, p. 103)].

(ix)

$$I(X) \leq I(X:Y) \leq I(X) + I(Y)$$

[Kolmogorov (1956)].

(x) I(X:X) = I(X) if X is discrete.

(xi) I(X:X) = I(X) = ∞ if X is continuous. This was proved by Corollary 4.2.1 and in Section 2.3.

## 4.5 Relationship between Singularity and Expected Mutual Information

(i) If $P_{XY}$ is not absolutely continuous with respect to $P_X \times P_Y$, as in Section 3.3 and in Theorem 4.2, then $I(X:Y) = \infty$. Equivalently, if $I(X:Y) < \infty$, then $P_{XY}$ is absolutely continuous with respect to $P_X \times P_Y$ [Gelfand and Yaglom (1959, p. 207)]. This is fairly obvious from Equation (4.3.1) since if $P_{XY}$ is not absolutely continuous with respect to $P_X \times P_Y$ then there exists a set A such that $P_X \times P_Y(A) = 0$ and $P_{XY}(A) \neq 0$. If we let A be one of the members of the partition (say $A = E_1$) then

$$I(X:Y) \geq P_{XY}(E_1) \log \frac{P_{XY}(E_1)}{0} = \infty$$

so that $I(X:Y) = \infty$.

(ii) If $P_{XY}$ is absolutely continuous with respect to $P_X \times P_Y$, then from Section 3.1.1,

$$I(X:Y) = E \log \frac{dP_{XY}}{d(P_X \times P_Y)} = \iint_{\Omega_T \times \Omega_S} \log \frac{dP_{XY}}{d(P_X \times P_Y)} \, dP_{XY} \,. \tag{4.5.1}$$

(iii) If $P_{XY}$ is absolutely continuous with respect to $P_X \times P_Y$ then we cannot conclude that $I(X:Y) < \infty$.

(iv) In Section 4.6.2 we shall see that if X and Y have multivariate normal distributions then $I(X:Y) = \infty$ if and only if $P_{XY} \perp P_X \times P_Y$ and $I(X:Y) < \infty$ if and only if $P_{XY} \parallel P_X \times P_Y$. We shall use the results of Section 3.3.3.

## 4.6 Expected Mutual Information for the Multivariate Normal Distributions

These results are from Pinsker (1964, pp. 122-127). Let $X' = (X_1, \ldots, X_n)$ and $Y' = (Y_1, \ldots, Y_n)$ be normal multivariate vectors with zero means. Define A to be the joint covariance matrix of X and Y. That is,

$$A = \begin{bmatrix} A_X & A_{XY} \\ \\ A_{YX} & A_Y \end{bmatrix} ,$$

where $A_X = E[XX']$, $A_{XY} = E[XY']$, $A_{YX} = E[YX']$, and $A_Y = E[YY']$. Let $|A|$ represent the determinant of $A$. We shall now consider the following cases.

(i) If $A \neq 0$ then $A_X \neq 0$ and $A_Y \neq 0$, and

$$I(X:Y) = \frac{1}{2} \log \frac{|A_X||A_Y|}{|A|} . \tag{4.6.0.1}$$

This follows from Section 4.1.1 since

$$I(X:Y) = h(X) + h(Y) - h(X,Y),$$

and

$$h(X) = \frac{1}{2} \log [(2\pi e)^n |A_X|] ,$$

$$h(Y) = \frac{1}{2} \log [(2\pi e)^m |A_Y|] ,$$

and

$$h(X,Y) = \frac{1}{2} \log [(2\pi e)^{m+n} |A|] ,$$

which can be shown similarly to Equation (4.6.3.4).

(ii) If $n = 1$ and $m = 1$ this reduces to

$$I(X:Y) = -\frac{1}{2} \log \left(1 - \frac{[E(XY)]^2}{E(X)^2 E(Y)^2}\right) = -\frac{1}{2} \log (1 - \rho^2) \tag{4.6.0.2}$$

or equivalently

$$\rho^2 = 1 - \exp[-2I(X:Y)]$$

where $\rho$ is the correlation coefficient of the variables X and Y [Linfoot (1957, p. 88)]. It can be seen directly in this case that $I(X:Y) = 0$ if and only if X and Y are independent ($\rho = 0$). $I(X:Y)$ is infinite if and only if $|\rho| = 1$ so that

$$P[(x,y) | x = ay+b] = 1$$

for some real constants a and b. [Cramér (1946, p. 265).]

(iii) If X and Y consist of linearly independent variables than a one-one linear transformation of **X** and **Y** can be made into $\mathbf{X}^t$ and $\mathbf{Y}^t$ so that all the $X_i^t$'s and $Y_j^t$'s are statistically independent with mean zero where $(i,j = 1, \ldots, k)$ and $k = \mathrm{Min}(n,m)$. That is

$$E[X_i^t X_j^t] = E[X_i^t Y_j^t] = E[Y_i^t Y_j^t] = 0 \quad \text{if } i \neq j .$$

Then $I(\mathbf{X}:\mathbf{Y})$ can be expressed as follows

$$I(\mathbf{X}:\mathbf{Y}) = I(\mathbf{X}^t:\mathbf{Y}^t) = -\frac{1}{2}\sum_{i=1}^{k} \log\left(1 - \frac{E^2[X_i^t Y_i^t]}{E(X_i^t)^2 E(Y_i^t)^2}\right) = -\frac{1}{2}\sum_{i=1}^{k}\log(1-\rho_i^2) \qquad (4.6.0.3)$$

where $k = \mathrm{Min}(n,m)$ and $\rho_i$ is the correlation coefficient of $X_i^t$ and $Y_i^t$. Here again it is seen that $I(\mathbf{X}:\mathbf{Y}) = 0$ if and only if $\mathbf{X}^t$ and $\mathbf{Y}^t$ are independent so that $\rho_i^2 = 0$ for all $(i = 1, \ldots, k)$. $I(\mathbf{X}:\mathbf{Y}) = \infty$ if and only if some $\rho_i^2 = 1$ so that $X_i^t$ is a linear function of $Y_i^t$ with probability one.

(iv) If $|\mathbf{A}_X||\mathbf{A}_Y| = 0$ then define $|\tilde{\mathbf{A}}_X|$ and $|\tilde{\mathbf{A}}_Y|$ to be any highest-order non-vanishing principal minors of $|\mathbf{A}_X|$ and $|\mathbf{A}_Y|$, and let $|\tilde{\mathbf{A}}|$ be the principal minor of $|\mathbf{A}|$ which contains $\tilde{\mathbf{A}}_X$ and $\tilde{\mathbf{A}}_Y$. If **A** is the zero matrix then define $\tilde{\mathbf{A}}$ also as the zero matrix. Then

$$\left.\begin{aligned} I(\mathbf{X}:\mathbf{Y}) &= \frac{1}{2}\log\frac{|\tilde{\mathbf{A}}_X||\tilde{\mathbf{A}}_Y|}{|\tilde{\mathbf{A}}|} \quad \text{if } |\tilde{\mathbf{A}}| \neq 0. \\ I(\mathbf{X}:\mathbf{Y}) &= \infty \text{ if } |\tilde{\mathbf{A}}| = 0. \end{aligned}\right\} \qquad (4.6.0.4)$$

### 4.6.1 Geometric Interpretation

(i) Define $B_Y$ as the set of random variables Z such that $P[Z = \sum_{i=1}^{m} c_i Y_i] = 1$ for fixed coefficients $c_1, \ldots, c_m$. Each component $X_i$ $(i = 1, \ldots, n)$ can be expressed as $X_i = \hat{X}_i + \check{X}_i$ where $\check{X}_i \in B_Y$ and $E[\hat{X}_i Z] = 0$ for all Z in $B_Y$ so that $E[\check{X}_i \hat{X}_i] = 0$.

(ii) Let $\hat{\mathbf{X}} = (\hat{X}_1, \ldots, \hat{X}_n)'$. If $|\tilde{\mathbf{A}}| \neq 0$ then

$$|\tilde{A}| = |\tilde{A}_Y||\tilde{A}_{\hat{X}}|$$

and

$$I(X:Y) = \frac{1}{2}\log\frac{|\tilde{A}_X||\tilde{A}_Y|}{|\tilde{A}|} = \frac{1}{2}\log\frac{|\tilde{A}_X|}{|\tilde{A}_{\hat{X}}|} \qquad (4.6.1.1)$$

if and only if $B_X$ and $B_Y$ do not have a common random variable. (That is, with probability 1 no $X_i$ is a linear combination of $Y_1, \ldots, Y_m$ and no $Y_i$ is a linear combination of $X_1, \ldots, X_n$.) Otherwise, $I(X:Y) = \infty$.

### 4.6.2 Relationship between Singularity and Expected Mutual Information for Multivariate Normal Vectors

We have the following theorem.

Theorem 4.6.2 If $X = (X_1, \ldots, X_n)'$ and $Y = (Y_1, \ldots, Y_m)'$ are multivariate normal vectors then $I(X:Y) = \infty$ if and only if $P_{XY} \perp P_X \times P_Y$ and $I(X:Y) < \infty$ if and only if $P_{XY} \parallel P_X \times P_Y$.

Proof:

(i) If $P_{XY} \perp P_X \times P_Y$ then $I(X:Y) = \infty$ from Section 4.5.

(ii) Therefore, if $I(X:Y) < \infty$ then $P_{XY}$ is not perpendicular to $P_X \times P_Y$. Then $P_{XY} \parallel P_X \times P_Y$ from Section 3.3.3.

(iii) If $|A| = 0$ then $P_{XY} \perp P_X \times P_Y$ from Section 3.3.3.

(iv) Therefore, if $P_{XY} \parallel P_X \times P_Y$ then $|A| \neq 0$ and $I(X:Y) < \infty$ from (4.6.0.1).

(v) Therefore, if $I(X:Y) = \infty$, then $P_{XY} \nparallel P_X \times P_Y$. Then $P_{XY} \perp P_X \times P_Y$ from Section 3.3.3.

### 4.6.3 I(R:S) between the Received and Original Signals

Let $R = S + N$ where $S$ and $N$ are independent n-dimensional normal column vectors with zero means and assume the covariance matrix of $N$ is non-singular. Now consider the set $B_S$ as defined in Section 4.6.1 (i). Each component $R_i$ $(i = 1, \ldots, n)$ can be expressed as $R_i = \hat{R}_i + \check{R}_i$ where $\check{R}_i = S_i$ and $\hat{R}_i = N_i$

because $S_i \in B_S$ and $E[S_i N_i] = 0$ $(i = 1, \ldots, n)$. It then follows from (4.6.1.1) that

$$I(\mathbf{R}:\mathbf{S}) = \frac{1}{2} \log \frac{|A_R|}{|A_{\hat{R}}|} = \frac{1}{2} \log \frac{|A_{S+N}|}{|A_N|} . \tag{4.6.3.0}$$

This is finite when $|A_N| \neq 0$. $B_R$ and $B_S$ cannot have a common random variable in this case for if they did, then for at least one $a_j \neq 0$ and $b_j \neq 0$,

$$P[\sum_{i=1}^{n} b_i S_i = \sum_{i=1}^{n} a_i (S_i + N_i)] = P[\sum_{i=1}^{n} (b_i - a_i) S_i = \sum_{i=1}^{n} a_i N_i]$$

$$= P\{N_j = \frac{1}{a_j} [\sum_{i=1}^{n} (b_i - a_i) S_i - \sum_{i \neq j}^{n} a_i N_i]\} = 1 .$$

This is impossible when $|A_N| \neq 0$ and when $\mathbf{S}$ and $\mathbf{N}$ are independent.

$A_{S+N}$ can be written as

$$A_{S+N} = E[(\mathbf{S} + \mathbf{N})(\mathbf{S} + \mathbf{N})'] = E(\mathbf{SS}') + E(\mathbf{SN}') + E(\mathbf{NS}') + E(\mathbf{NN}')$$

$$= E(\mathbf{SS}') + E(\mathbf{NN}') = A_S + A_N , \tag{4.6.3.1}$$

because $\mathbf{S}$ and $\mathbf{N}$ are independent. Therefore, $I(\mathbf{R}:\mathbf{S})$ can be expressed as

$$I(\mathbf{R}:\mathbf{S}) = \frac{1}{2} \log \frac{|A_S + A_N|}{|A_N|} = \frac{1}{2} \log |I + A_S A_N^{-1}| . \tag{4.6.3.2}$$

If $A_S = \alpha A_N$, then

$$I(\mathbf{R}:\mathbf{S}) = \frac{1}{2} \log |(1 + \alpha) I| = \frac{n}{2} \log (1 + \alpha). \tag{4.6.3.3}$$

The results of this section may also be found by using the result of Section 4.1.3 as shown by the following argument which is due to Good and Doog (1958).

In Section 4.1.3 it was found that $I(\mathbf{R}:\mathbf{S}) = h(\mathbf{R}) - h(\mathbf{N})$ where $\mathbf{R}$, $\mathbf{S}$, and $\mathbf{N}$ are now n-dimensional column vectors.

$$h(\mathbf{R}) = -E[\log f(R_1, \ldots, R_n)]$$

$$= -E\{\log\{[(2\pi)^{(n|2)}|\mathbf{A_R}|^{\frac{1}{2}}]^{-1}\exp(-\frac{1}{2}\,\mathbf{R}'\mathbf{A_R}^{-1}\mathbf{R})\}\}$$

$$= E\{\frac{1}{2}\log[(2\pi)^n|\mathbf{A_R}|] + \frac{1}{2}\chi^2 \text{ (chi-squared)}\}$$

$$= \frac{1}{2}\log[(2\pi)^n|\mathbf{A_R}|] + \frac{n}{2}$$

$$= \frac{1}{2}\log[(2\pi e)^n|\mathbf{A_R}|]\,. \qquad (4.6.3.4)$$

Similarly,

$$h(\mathbf{N}) = \frac{1}{2}\log[(2\pi e)^n|\mathbf{A_N}|] \qquad (4.6.3.5)$$

and

$$I(R:S) = h(R) - h(N) = \frac{1}{2}\log\frac{(2\pi e)^n|\mathbf{A_R}|}{(2\pi e)^n|\mathbf{A_N}|}$$

$$= \frac{1}{2}\log\frac{|\mathbf{A_S}+\mathbf{A_N}|}{|\mathbf{A_N}|} = \frac{1}{2}\log|\mathbf{I} + \mathbf{A_S}\mathbf{A_N}^{-1}| \qquad (4.6.3.6)$$

which agrees with (4.6.3.2).

### 4.6.4 Expected Mutual Information for Infinite Sets of Normal Variables

Let $\mathbf{X} = (X_1, X_2, \ldots)$ and $\mathbf{Y} = \{Y(s)|s \in S\}$ where S is any arbitrary set. Assume that all finite subsets of $\mathbf{X}$ and $\mathbf{Y}$ have multivariate normal distributions. The results of Section 4.6.1 can be extended to fit this case. Let $B_{\mathbf{Y}}$ be the set of all random Z such that

$$P[Z = \lim_{m\to\infty}\sum_{i=1}^{m} c_i\,Y(s_i)] = 1$$

for some set of constants $c_i$ and $s_i \in S$ $(i = 1, 2, \ldots)$. Let $(X_1, \ldots, X_n)$ $(n = 1, 2, \ldots)$ be the first n components of $\mathbf{X}$. Then (4.6.1.1) can be extended so that

$$I(\mathbf{X}:\mathbf{Y}) = \lim_{n\to\infty} I[(X_1,\ldots, X_n):\mathbf{Y}] = \lim_{n\to\infty}\frac{1}{2}\log\frac{|\tilde{\mathbf{A}}_{\mathbf{X}}|}{|\tilde{\mathbf{A}}_{\hat{\mathbf{X}}}|} \qquad (4.6.4.1)$$

if and only if $B_X$ and $B_Y$ to not have a common random variable. If they do then $I(X:Y) = \infty$. If $X = X_1$ is one-dimensional then

$$I(X_1:Y) = \frac{1}{2} \log \frac{E(X_1^2)}{E(\hat{X}_1^2)}$$

if and only if $X_1 \notin B_Y$. If $X_1 \in B_Y$ then $I(X_1:Y) = \infty$.

## V. EXPECTED WEIGHT OF EVIDENCE

The expected weight of evidence in favor of one hypothesis $H_1$ compared with another, $H_2$, given $H_1$ is true, when the observation is a random vector, is discussed in this chapter. The expected weight of evidence is shown to be infinite when the probability measure of a random vector, given $H_1$ is true, is not absolutely continuous with respect to the probability measure given that $H_2$ is true. The special case of normal vectors is discussed.

### 5.1 General Expressions

Let $P_1$ and $P_2$ be two probability measures defined on the same measurable space $(\Omega,A)$ under two opposing hypotheses $H_1$ and $H_2$. Let $E_i$ $(i = 1, 2, \ldots)$ be a partition of $\Omega$, so that $\Omega = \bigcup_{i=1}^{\infty} E_i$, where the $E_i$ are mutually disjoint. Define the expected weight of evidence in favor of $H_1$ compared to $H_2$ given $H_1$ as

$$W(H_1/H_2) = W(H_1/H_2|H_1) = \sup \sum_{i=1}^{\infty} P_1(E_i)\log\frac{P_1(E_i)}{P_2(E_i)} \tag{5.1.1}$$

where the supremum is taken over all partitions of $\Omega$. If $P_1 = P_{\mathbf{XY}}$ and $P_2 = P_{\mathbf{X}} \times P_{\mathbf{Y}}$ as in Section 4.3 then $I(\mathbf{X}:\mathbf{Y}) = W(H_1/H_2)$ so that $I(\mathbf{X}:\mathbf{Y})$ is a special case of $W(H_1/H_2)$. It should be observed that $W(H_1/H_2)$ does not have the same meaning as $W(H_1/H_2:B)$ did in Section 1.4.2.

If $W(H_1/H_2) < \infty$ then $P_1$ is absolutely continuous with respect to $P_2$ and

$$W(H_1/H_2) = \int_{\Omega} \log\frac{dP_1}{dP_2}\,dP_1 = E[\log\frac{dP_1}{dP_2}|\,H_1] \tag{5.1.2}$$

which may be abbreviated as

$$W(H_1/H_2) = E_1 \log\frac{dP_1}{dP_2}\,.$$

If $\mathbf{X}$ is a discrete random vector then

$$\frac{dP_1(\mathbf{x})}{dP_2(\mathbf{x})} = \frac{P[\mathbf{X}=\mathbf{x}|H_1]}{P[\mathbf{X}=\mathbf{x}|H_2]} = \frac{p_1(\mathbf{x})}{p_2(\mathbf{x})} \tag{5.1.3}$$

where $p_1$ and $p_2$ are the discrete probability mass functions. If $\mathbf{X}$ has a continuous probability density function $f_1(\mathbf{x})$ and $f_2(\mathbf{x})$ under the two opposing hypotheses $H_1$ and $H_2$, then

$$\frac{dP_1(\mathbf{x})}{dP_2(\mathbf{x})} = \frac{f_1(\mathbf{x})}{f_2(\mathbf{x})} .$$

Most of the ideas of Section 5.1 were from Pinsker (1964, p. 19).

In the following example $P_1$ is not absolutely continuous with respect to $P_2$ and so $W(H_1/H_2) = \infty$.

## 5.2 Discrete Singular Uniform Probabilities

In the example in Section 3.2.1 set $S_1 = \{1,2\}$ and $S_2 = \{3\}$. Let $p_1(x) = \frac{1}{3}$ if $x = 1,2$, or 3, and let $p_2(x) = \frac{1}{2}$ if $x = 1$ or 2. Then $\Omega = S_1 \cup S_2$. Therefore, from the definition (5.1.1),

$$W(H_1/H_2) \geq P_1(S_1)\log\frac{P_1(S_1)}{P_2(S_1)} + P_1(S_2)\log\frac{P_1(S_2)}{P_2(S_2)}$$

$$= \frac{2}{3}\log\frac{(2/3)}{1} + \frac{1}{3}\log\frac{(1/3)}{0} = \infty.$$

$P_1$ is not absolutely continuous with respect to $P_2$ since $P_2(S_2) = 0$, but $P_1(S_2) = \frac{1}{3}$.

## 5.3 Properties of Expected Weight of Evidence

If

$$E_1 \log\frac{dP_1(\mathbf{x},\mathbf{y})}{dP_2(\mathbf{x},\mathbf{y})}$$

is denoted by $W(H_1/H_2{:}\mathbf{X},\mathbf{Y})$,

$$E_1 \log\frac{dP_1(\mathbf{x})}{dP_2(\mathbf{x})}$$

is denoted by $W(H_1/H_2:X)$, and

$$E_1 \log \frac{dP_1(y|x)}{dP_2(y|x)}$$

is denoted by $W(H_1/H_2:Y|X)$, then the following properties are shown, for example, by Kullback (1959, pp. 12-22).

(i) $W(H_1/H_2:X,Y) = W(H_1/H_2:X) + W(H_1/H_2:Y|X) = W(H_1/H_2:Y) + W(H_1/H_2:X|Y)$.

(ii) If X and Y are independent then

$$W(H_1/H_2:X,Y) = W(H_1/H_2:X) + W(H_1/H_2:Y) .$$

(iii) $W(H_1/H_2) \geq 0$, with equality if and only if $\frac{dP_1}{dP_2} = 1$ over $\Omega$. Good (1950, pp. 72 and 75) attributes this property to Turing (1940).

(iv) If $Y = T(X)$ is a one-one transformation, then $W(H_1/H_2:X) \geq W(H_1/H_2:Y)$ with equality if and only if $\frac{dP_1(x)}{dP_2(x)} = \frac{d\bar{P}_1(y)}{d\bar{P}_2(y)}$ where $\bar{P}_1$ and $\bar{P}_2$ are the probability measures associated with Y.

## 5.4 Various Names for Expected Weight of Evidence

(i) Peirce (1878), Good (1950, p. 75), and Minsky and Selfridge (1961) independently imply that it should be called the expected weight of evidence.

(ii) Kullback (1959, p. 5) names it the mean information.

(iii) Yaglom (1962, p. 333) names it the entropy distance.

(iv) Pinsker (1964, p. 19) names it the entropy.

(v) Good (1969a, p. 198) names it dinegentropy.

## 5.5 Various Interpretations of the Expected Weight of Evidence when the Variable is Discrete

If X is a discrete random variable then from Equations (5.1.2) and (5.1.3), and Section 1.2,

$$\begin{aligned} W(H_1/H_2) &= E_1\{\log P[X = x|H_1] - \log P[X = x|H_2]\} \\ &= E_1[I(x|H_2) - I(x|H_1)]. \end{aligned} \tag{5.5.1}$$

Therefore,

$$W(H_1/H_2) = E_1[I(x:H_1) - I(x:H_2)] \tag{5.5.2}$$

from Equation (1.4.2.2). This may be interpreted as the expected additional information about x provided by $H_1$ as opposed to $H_2$, given $H_1$.

Now, from (1.4.1.2), $W(H_1/H_2)$ can also be expressed as

$$\begin{aligned} W(H_1/H_2) &= E_1[I(H_1:x) - I(H_2:x)] \\ &= E_1\left[\log \frac{P[H_1|x]}{P[H_1]} - \log \frac{P[H_2|x]}{P[H_2]}\right] \\ &= E_1\left[\log \frac{P[H_1|x]}{P[H_2|x]} - \log \frac{P[H_1]}{P[H_2]}\right] \\ &= E_1\left[\log \frac{O(H_1/H_2|x)}{O(H_1/H_2)}\right] \end{aligned} \tag{5.5.3}$$

where $O(H_1/H_2|x)$ is the odds in favor of $H_1$ as opposed to $H_2$, given $X = x$, and $O(H_1/H_2)$ is the odds in favor of $H_1$ as opposed to $H_2$. $H_1$ and $H_2$ may be thought of as events in the <u>a priori</u> probability spaces $(\Omega, \mathcal{H}, P)$ and $(\Omega, \mathcal{H}, P_x)$ where $P_x$ is the conditional probability given $X = x$. $W(H_1/H_2)$ may be interpreted as the expected additional information about $H_1$ as opposed to $H_2$ provided by x given $H_1$.

## 5.6 $W(H_{S+N}/H_N)$ for Multivariate Normal Vectors

Let $\mathbf{R}_1 = \mathbf{S} + \mathbf{N}$ and $\mathbf{R}_2 = \mathbf{N}$, where $\mathbf{S}$ and $\mathbf{N}$ are independent n-dimensional normally distributed column vectors with zero means and assume, as in Section 4.6.3, that $|\mathbf{A}_N| \neq 0$. If $H_{S+N}$ is the hypothesis that $\mathbf{R} = \mathbf{R}_1 = \mathbf{S} + \mathbf{N}$ and $H_N$ is the hypothesis that $\mathbf{R} = \mathbf{R}_2 = \mathbf{N}$, then from (5.1.2)

$$W(H_{S+N}/H_N) = E_{S+N} \log \frac{f_{S+N}(R)}{f_N(R)}$$

$$= E_{S+N} \log \left\{ \frac{[(2\pi)^{\frac{n}{2}} |A_S+A_N|^{\frac{1}{2}}]^{-1} \exp[-\frac{1}{2} R'(A_S+A_N)^{-1}R]}{[(2\pi)^{\frac{n}{2}} |A_N|^{\frac{1}{2}}]^{-1} \exp[-\frac{1}{2} R' A_N^{-1} R]} \right\}$$

$$= E_{S+N}\{-\frac{1}{2}\log|I+A_SA_N^{-1}| + \frac{1}{2}R'[A_N^{-1} - (A_S+A_N)^{-1}]R\}$$

$$= -\frac{1}{2}\log|I+A_SA_N^{-1}| + \frac{1}{2}E_{S+N}\{\mathrm{tr}\, R'[A_N^{-1} - (A_S+A_N)^{-1}]R\}$$

because the trace of the scalar $R'[\ldots]R$ is always equal to itself (tr denotes the trace.) Therefore

$$W(H_{S+N}/H_N) = -\frac{1}{2}\log|I+A_SA_N^{-1}| + \frac{1}{2}\mathrm{tr}[A_N^{-1}A_S + I - I]$$

$$= -\frac{1}{2}\log|I+A_SA_N^{-1}| + \frac{1}{2}\mathrm{tr}[A_SA_N^{-1}]. \qquad (5.6.1)$$

[For example, Good (1960a, p. 119).]

## 5.7 Alternate Form for Expected Weight of Evidence

If we define

$$W(H_1/H_2|H_2) = \sup \sum_{i=1}^{\infty} P_2(E_i) \log \frac{P_1(E_i)}{P_2(E_i)}$$

$$= -W(H_2/H_1|H_2) \le 0 \qquad (5.7.1)$$

similarly to (5.1.1), then all the properties that hold for $W(H_1/H_2) = W(H_1/H_2|H_1)$ are like the properties that hold for $W(H_1/H_2|H_2)$.

Since $E_N(RR') = E(NN') = A_N$, it follows from Section 5.6 that

$$W(H_{S+N}/H_N|H_N) = -\frac{1}{2}\log|I+A_SA_N^{-1}| + \frac{1}{2}\mathrm{tr}\{[A_N^{-1}-(A_S+A_N)^{-1}]A_N\}$$

$$= -\frac{1}{2}\log|I+A_SA_N^{-1}| + \frac{1}{2}\mathrm{tr}[I-(A_N^{-1}A_S+I)^{-1}]\ . \qquad (5.7.2)$$

# VI. DIVERGENCE

Divergence is discussed in this chapter. It is shown to be infinite when the probability measures under the two alternate hypotheses are not parallel. The special case of normal vectors is considered. Most of the results of this chapter are from Kullback (1959, pp. 6 and 22-26). Divergence was previously used in classified papers by Good and later appeared in Jeffreys (1946) and (1948, p. 158).

## 6.1 General Expression

The divergence between the hypotheses $H_1$ and $H_2$ is defined as

$$J(1,2) = W(H_1/H_2|H_1) - W(H_1/H_2|H_2) \tag{6.1.1}$$

$$= W(H_2/H_1|H_2) - W(H_2/H_1|H_1) \tag{6.1.2}$$

$$= W(H_1/H_2|H_1) + W(H_2/H_1|H_2) \tag{6.1.3}$$

$$= W(H_1/H_2) + W(H_2/H_1) \ . \tag{6.1.4}$$

It is symmetric with respect to $H_1$ and $H_2$ and (6.1.4) may be interpreted as the expected weight of evidence in favor of $H_1$ as opposed to $H_2$, given $H_1$, plus the expected weight of evidence in favor of $H_2$ as opposed to $H_1$, given $H_2$. The divergence may also be directly defined as

$$J(1,2) = \sup \sum_{i=1}^{\infty} [P_1(E_i) - P_2(E_i)] \log \frac{P_1(E_i)}{P_2(E_i)} \tag{6.1.5}$$

where the supremum is taken over all possible partitions of $\Omega$.

## 6.2 Relationship between Parallelism of $P_1$ and $P_2$ and Divergence

The following theorem is assumed in Kullback (1959, p. 6) although not proved explicitly.

Theorem 6.2. If $J(1,2) < \infty$ then $P_1$ is parallel to $P_2$ (which means that

$P_1$ and $P_2$ are absolutely continuous with respect to each other).

Proof: Suppose $P_1 \not\parallel P_2$. If $P_1$ is not absolutely continuous with respect to $P_2$ then there exists a set $S_1$ such that $P_1(S_1) > 0$ and $P_2(S_1) = 0$. Hence, from (6.1.5)

$$J(1,2) \geq (P_1(S_1) - 0) \log \left(\frac{P_1(S_1)}{0}\right) = \infty.$$

If $P_2$ is not absolutely continuous with respect to $P_1$, then there is a set $S_2$ such that $P_2(S_2) > 0$ and $P_1(S_2) = 0$. Thus,

$$J(1,2) \geq [0 - P_2(S_2)] \log \left(\frac{0}{P_2(S_2)}\right) = \infty.$$

Therefore, if $J(1,2) < \infty$, then $P_1 \parallel P_2$. Q.E.D.

If $J(1,2) < \infty$, then $J(1,2)$ is expressed as

$$J(1,2) = \int_\Omega (dP_1 - dP_2) \log \frac{dP_1}{dP_2} \tag{6.2.1}$$

where $P_1$ and $P_2$ are two probability measures defined on the same measurable space $(\Omega, A)$. (See Section 3.1.1 for an explanation of the notations.)

## 6.3 Various Other Expressions for Divergence between Parallel Probability Measures

If $P_1 \parallel P_2$, then $J(1,2)$ can be expressed in the following manner.

(i) $J(1,2) = \int_\Omega \left(\frac{dP_1}{dP_2} - 1\right)\left(\log \frac{dP_1}{dP_2}\right) dP_2.$

This is similar to the definition in Hájek (1958a, p. 460).

(ii) If the variable is discrete then from (5.5.3) and (6.1.1) it follows that

$$J(1,2) = \sum_{i=1}^{\infty} (p_1(x_i) - p_2(x_i)) \log O(H_1/H_2 | x_i) .$$

This expression is similar to the definition in Jeffreys (1948, p. 58).

## 6.4 Properties of Divergence

The following properties are listed in Kullback (1959, pp. 22-26).

(i) $J(1,2) \geq 0$ with equality if and only if $P_1$ and $P_2$ are identical. This follows immediately from (5.3(iii)) and (6.1.4).

(ii) Using notation similar to Section 5.3, the following relationship holds:

$$J(1,2:X,Y) = J(1,2:X) + J(1,2:Y|X) = J(1,2:Y) + J(1,2:X|Y) = J(1,2:X) + J(1,2:Y)$$

if X and Y are independent.

(iii) If $Y = T(X)$ is a linear transformation then $J(1,2:X) \geq J(1,2;Y)$ with equality if and only if Y is a nonsingular linear transformation.

(iv) If $Y = T(X)$ is any transformation and $g(Y) = g(T(X)) = h(X)$ is an arbitrary function of Y, then

$$\int_\Omega h(\mathbf{x})[dP_1(\mathbf{x}) - dP_2(\mathbf{x})] \log \left(\frac{dP_1(\mathbf{x})}{dP_2(\mathbf{x})}\right) \geq \int_\Omega g(\mathbf{y})[dP_1^*(\mathbf{y}) - dP_2^*(\mathbf{y})] \log \frac{dP_1^*(\mathbf{y})}{dP_2^*(\mathbf{y})}$$

with equality if and only if $Y = T(X)$ is a one-one transformation.

## 6.5 J(S + N,N) for Multivariate Normal Vectors

Let $R_1 = S + N$ and $R_2 = N$, where S and N are defined as in Section 5.6. Then it follows from (5.6.1) and (5.7.2) that

$$J(S+N,N) = W(H_{S+N}/H_N|H_{S+N}) - W(H_{S+N}/H_N|H_N) = -\frac{1}{2}\log|I+A_SA_N^{-1}| + \frac{1}{2}\mathrm{tr}A_SA_N^{-1}$$

$$+ \frac{1}{2}\log|I+A_SA_N^{-1}| - \frac{1}{2}\mathrm{tr}[I-(A_N^{-1}A_S+I)^{-1}]$$

$$= \frac{1}{2}\mathrm{tr}(A_SA_N^{-1}) - \frac{1}{2}\mathrm{tr}[I-(I+A_N^{-1}A_S)^{-1}]. \qquad (6.5.1)$$

# VII. EXPECTED MUTUAL INFORMATION, EXPECTED WEIGHT OF EVIDENCE, AND DIVERGENCE FOR RANDOM PROCESSES

In this chapter expected mutual information, expected weight of evidence, and divergence are discussed for random processes. Their relationships to singular and parallel probability measures, errorless discrimination between two random processes, and errorless signal detection over arbitrarily small intervals are also discussed.

## 7.1 Joint and Product Probability Measures in Function Space

Let $\mathbf{X}_T = \{X(t)|t \text{ in } T\}$ and $\mathbf{Y}_S = \{Y(s)|s \text{ in } S\}$ be families of random variables, where T and S represent the intervals [0,T] and [0,S], respectively, as in Parzen (1962a, p. 168). $\mathbf{X}_T$ and $\mathbf{Y}_S$ are then said to be stochastic (or random) processes on T and S.

Let $(\Omega_T, \mathcal{B}_T, P_{\mathbf{X}_T})$ and $(\Omega_S, \mathcal{B}_S, P_{\mathbf{Y}_S})$ represent the probability spaces, where $\Omega_T$ and $\Omega_S$ are the sets of all real-valued functions on T and S and $\mathcal{B}_T$ and $\mathcal{B}_S$ are the Borel $\sigma$-fields defined on subsets of $\Omega_T$ and $\Omega_S$, respectively.

The joint probability $P_{\mathbf{X}_T\mathbf{Y}_S}$ and the product probability $P_{\mathbf{X}_T} \times P_{\mathbf{Y}_S}$ are defined on the measurable space $(\Omega_T \times \Omega_S, \mathcal{B}_T \times \mathcal{B}_S)$ where $\Omega_T \times \Omega_S$ is the set of all real-valued functions on T × S and $\mathcal{B}_T \times \mathcal{B}_S$ is the $\sigma$-field defined for sets in $\Omega_T \times \Omega_S$. $P_{\mathbf{X}_T} \times P_{\mathbf{Y}_S}$ has the property that

$$P_{\mathbf{X}_T} \times P_{\mathbf{Y}_S}(B_1 \times B_2) = P_{\mathbf{X}_T}(B_1)\, P_{\mathbf{Y}_S}(B_2)$$

for all $B_1$ in $\mathcal{B}_T$ and $B_2$ in $\mathcal{B}_S$. If B is any set in $\mathcal{B}_T \times \mathcal{B}_S$ then $P_{\mathbf{X}_T} \times P_{\mathbf{Y}_S}(B)$ may be found by using the following form of Fubini's theorem from Loève (1963, p. 135).

$$P_{\mathbf{X}_T} \times P_{\mathbf{Y}_S}(B) = \int_B d(P_{\mathbf{X}_T} \times P_{\mathbf{Y}_S}) = \int_{\mathbf{y}} dP_{\mathbf{Y}_S} \int_{B_{\mathbf{y}}} dP_{\mathbf{X}_T} = \int_{\mathbf{y}} P_{\mathbf{X}_T}(B_{\mathbf{y}})\, dP_{\mathbf{Y}_S}$$

where $B_{\mathbf{y}} = \{\mathbf{x} | (\mathbf{x},\mathbf{y}) \text{ in } B\}$ is a set function of **x** given **y**.

## 7.2 Expected Mutual Information

It follows from Definition (4.5.1) that

$$I(\mathbf{X}_T:\mathbf{Y}_S) = \iint_{\Omega_T \times \Omega_S} \log \frac{dP_{\mathbf{X}_T\mathbf{Y}_S}}{dP_{\mathbf{X}_T} \times P_{\mathbf{Y}_S}} \, dP_{\mathbf{X}_T\mathbf{Y}_S}$$

if $P_{\mathbf{X}_T\mathbf{Y}_S}$ is absolutely continuous with respect to $P_{\mathbf{X}_T} \times P_{\mathbf{Y}_S}$. Otherwise $I(\mathbf{X}_T:\mathbf{Y}_S) = \infty$.

$I(\mathbf{X}_T:\mathbf{Y}_S)$ may also be defined by first sampling at the points $T_n = \{t_i\}$ $(i = 1, \ldots, n)$ where $T_n \subseteq T$, and $S_m = \{s_j\}$ $(j = 1, \ldots, m)$ where $S_m \subseteq S$, to obtain $\mathbf{X}_n = \{X(t_i)\}$ ($t_i$ in $T_n$) and $\mathbf{Y}_m = \{Y(s_i)\}$ ($s_i$ in $S_m$) which represent ordered samples from $\mathbf{X}_T$ and $\mathbf{Y}_S$, respectively. If $P_{\mathbf{X}_n\mathbf{Y}_m}$ is absolutely continuous with respect to $P_{\mathbf{X}_n} \times P_{\mathbf{Y}_m}$ then

$$I(\mathbf{X}_n:\mathbf{Y}_m) = \iint_{\Omega_{T_n} \times \Omega_{S_m}} \log \frac{dP_{\mathbf{X}_n\mathbf{Y}_m}}{dP_{\mathbf{X}_n} \times P_{\mathbf{Y}_m}} \, dP_{\mathbf{X}_n\mathbf{Y}_m} .$$

Otherwise $I(\mathbf{X}_n:\mathbf{Y}_m) = \infty$.

Gelfand and Yaglom (1959, p. 206) proved that the addition of new points can only increase $I(\mathbf{X}_n:\mathbf{Y}_m)$. Therefore, Yaglom (1962, p. 332) and Good (1958, p. 115) define

$$I(\mathbf{X}_T:\mathbf{Y}_S) = \sup_{\substack{T_n, S_m \\ n,m}} I(\mathbf{X}_n:\mathbf{Y}_m) = \lim_{\substack{T_n \to T \\ S_m \to S}} I(\mathbf{X}_n:\mathbf{Y}_m)$$

where $T_n \to T$ is the Riemannian limit [which means that $n \to \infty$ and $\mathrm{Max}|t_i - t_j| \to 0$ for all $(i = 1, 2, \ldots)$ and $(j = 1, 2, \ldots)$].

## 7.3 Two Random Processes over the Same Interval

In Sections 7.4 and 7.5 we shall consider two random processes

$X_{1T} = \{X_1(t)\}$ (t in T) and $X_{2T} = \{X_2(t)\}$ (t in T) under two opposing hypotheses $H_1$ and $H_2$, where, to avoid the introduction of a new symbol, we denote the interval $[0,T]$ by T. Let $P_{1T}$ and $P_{2T}$ represent the probability measures of $X_{1T}$ and $X_{2T}$ defined on the measurable space $(\Omega_T, A_T)$, where $\Omega_T$ is the set of all real-valued functions on T, and $A_T$ is a Borel $\sigma$-field of subsets of $\Omega_T$.

## 7.4 Expected Weight of Evidence

It follows from (5.1.2) that

$$W_T(H_1/H_2) = \int_{\Omega_T} \log \left(\frac{dP_{1T}}{dP_{2T}}\right) dP_{1T}$$

if $P_{1T}$ is absolutely continuous with respect to $P_{2T}$. Otherwise $W_T(H_1/H_2) = \infty$.

$W_T(H_1/H_2)$ may also be defined by first sampling at the points $T_n = \{t_i\}$ $(i = 1, 2, \ldots, n)$ where $T_n \subseteq T$ to obtain $X_{1n} = \{X_1(t_i)\}$ and $X_{2n} = \{X_2(t_i)\}$ ($t_i$ in $T_n$) which represent ordered samples from $X_1$ and $X_2$. Let $P_{1n}$ and $P_{2n}$ represent the probability measures of $X_{1n}$ and $X_{2n}$. If $P_{1n}$ is absolutely continuous with respect to $P_{2n}$ then

$$W_n(H_1/H_2) = \int_{\Omega_{T_n}} \log \left(\frac{dP_{1n}}{dP_{2n}}\right) dP_{1n}.$$

Otherwise $W_n(H_1/H_2) = \infty$.

Hájek (1958a) proved (as one would intuitively demand) that the addition of new points can only increase $W_n(H_1/H_2)$ and Yaglom (1962, p. 333) concludes that

$$W_T(H_1/H_2) = \sup_{n,T_n} W_n(H_1/H_2) = \lim_{T_n \to T} W_n(H_1/H_2)$$

where $T_n \to T$ denotes the Riemannian limit as defined in Section 7.2.

## 7.5 Divergence

If $P_{1n} \,||\, P_{2n}$, then

$$J_n(1,2) = \int_{\Omega_{T_n}} (dP_{1n} - dP_{2n}) \log \frac{dP_{1n}}{dP_{2n}} ;$$

otherwise $J_n(1,2) = \infty$. Hájek (1958a) defined

$$J_T(1,2) = \sup_{T_n, n} J_n(1,2) = \lim_{T_n \to T} J_n(1,2)$$

where $T_n \to T$ denotes the Riemannian limit. If $J_T(1,2)$ is finite then $P_{1T} \parallel P_{2T}$ and

$$J_T(1,2) = \int_{\Omega_T} (dP_{1T} - dP_{2T}) \log \frac{dP_{1T}}{dP_{2T}} .$$

If $P_{1T} \nparallel P_{2T}$ then $J_T(1,2) = \infty$.

## 7.6 Gaussian Processes

A random process $\mathbf{X}_T = \{X(t)\}$ ($t$ in $T$), where $T$ is <u>any</u> ordered set, is called a <u>gaussian process</u> if $\mathbf{X}_n = \{x(t_i)\}$ ($t_i$ in $T$, $i = 1, \ldots, n$) has a multivariate normal distribution for all finite subsets $T_n = \{t_i\}$ ($i = 1, \ldots, n$). Two of the reasons why gaussian processes are important are given below. Another is given in Section 12.6.1. [See also Good (1966a, p. 381) and Kolmogorov (1956, p. 106).]

(i) Gaussian processes often arise as an adequate approximation when one has a sum of processes emitted by a large number of independent sources.

For instance, noise in electrical circuits is usually approximately gaussian, as are atmospheric turbulence, tidal heights, and fluctuations in many dynamical systems.

(ii) Gaussian covariance stationary processes with zero mean are completely determined by the covariance function or the spectral density function, and are strictly stationary, provided they are covariance stationary. (See Section 8.1 for the definitions of these terms.)

This summary is from Breiman (1969).

## 7.7 Relationship between Divergence and the Singularity between Gaussian Probability Measures

Hájek (1958b) has shown that for gaussian stochastic processes the following relationship holds.

Theorem 7.7.1. For two gaussian processes, if $J_T(1,2) = \infty$ then $P_{1T} \perp P_{2T}$ even though $P_{1n} \parallel P_{2n}$ and $J_n(1,2) < \infty$ for all finite n.

Hájek first proved the following Lemma which is essential for this main result.

Lemma 7.7.1. Let $X_{1n}$ and $X_{2n}$ have multivariate normal distributions (where $X_{1n}$ and $X_{2n}$ are defined in Section 7.4). Then for all $\varepsilon > 0$, there exists a constant $K_\varepsilon$ such that if $J_n(1,2) > K_\varepsilon$, then there exists a set $B_n$ such that $P_{1n}(B_n) < \varepsilon$ and $P_{2n}(B_n) > 1 - \varepsilon$.

Proof of Theorem 7.7.1. Suppose that $J_T(1,2) = \infty$. For any integer $m > 0$, let $\varepsilon = 2^{-m} > 0$. For any constant $K_\varepsilon$, there exists an integer $n_m$ such that for all $n' > n_m$ there is a collection of $n'$ points $t_i$ $(i = 1, 2, \ldots, n')$ such that $J_{n'}(1,2) > K_\varepsilon$. Therefore, from the lemma there exists a set $B_{n'}$ such that $P_{1n'}(B_{n'}) < 2^{-m}$ and $P_{2n'}(B_{n'}) > 1 - 2^{-m}$.

Therefore, since $(\Omega_{T_{n'}}, \mathcal{B}_{T_{n'}}, P_{j_{n'}}) \underset{T_{n'}\to T}{\longrightarrow} (\Omega_T, \mathcal{B}_T, P_{jT})$ $(j = 1,2)$, there exists a set $A_m$ in $\mathcal{B}_T$ such that $P_{1T}(A_m) < 2^{-m}$ and $P_{2T}(A_m) > 1 - 2^{-m}$ for $m = 1, 2, \ldots$. Let $A = \bigcap_{k=1}^{\infty} \bigcup_{m=k}^{\infty} A_m \subseteq \mathcal{B}_T$ since $\mathcal{B}_T$ is a $\sigma$-field. Then

$$P_{1T}(A) = P_{1T}\Big(\bigcap_{k=1}^{\infty} \bigcup_{m=k}^{\infty} A_m\Big) = P_{1T}\Big(\lim_{k\to\infty} \bigcup_{m=k}^{\infty} A_m\Big) = \lim_{k\to\infty} P_{1T}\Big(\bigcup_{m=k}^{\infty} A_m\Big)$$

from Loève (1963, p. 150). Therefore,

$$P_{1T}(A) \le \lim_{k\to\infty} \sum_{m=k}^{\infty} P_{1T}(A_m) < \lim_{k\to\infty} \sum_{m=k}^{\infty} 2^{-m}$$

$$= \lim_{k\to\infty} 1 - (\tfrac{1}{2}) \frac{1-(\tfrac{1}{2})^k}{1-\tfrac{1}{2}} = \lim_{k\to\infty} (\tfrac{1}{2})^k = 0.$$

Since $P_{1T}(A) \ge 0$ it follows that $P_{1T}(A) = 0$. Similarly, it can be shown that $P_{2T}(A) = 1$ so that $P_{1T} \perp P_{2T}$. Q.E.D.

In Section 7.5 it was shown that, given any two stochastic processes $X_{1T}$ and $X_{2T}$, we have $J_T(1,2) < \infty$ implies $P_{1T} \parallel P_{2T}$. Theorem 7.7.1 states that if $X_{1T}$ and $X_{2T}$ are gaussian processes then $J_T(1,2) = \infty$ implies $P_{1T} \perp P_{2T}$. Therefore, it can be concluded that the probability measures of gaussian processes are either parallel or perpendicular depending on whether $J_T(1,2)$ is finite or infinite.

## 7.8 Relationship between Expected Weight of Evidence, Mutual Information, and Singularity of Gaussian Probability Measures

Yaglom (1962, p. 333) states that the probability measures of gaussian processes are either parallel or perpendicular depending on whether $W_T(H_1/H_2)$ is finite or infinite. This follows from Section 7.7, Equation (6.1.4), and Section 5.3(iii). Yaglom (p. 332) also states that the probability measures $P_{X_T Y_S}$ and $P_{X_T} \times P_{Y_S}$ are either parallel or perpendicular depending on whether $I(X_T:Y_S)$ is finite or infinite when $X_T$ and $Y_S$ are gaussian processes. Theorem 4.6.2 is a special case of this result when T is a finite set.

## 7.9 Relationship between Perpendicular Probability Measures of Random Processes and Perfect Discrimination

Suppose the probability measures of two random processes defined over an arbitrarily small interval T under two opposing hypotheses are perpendicular. Grenander (1950) and Yaglom (1952, p. 339) show that this assumption of perpendicularity is equivalent to the possibility of discrimination with arbitrarily small error between the two hypotheses by only observing the values of these processes on the arbitrarily small interval T.

For two gaussian processes the above situation would always occur when $J_T(1,2)$ and $W_T(H_1/H_2)$ are infinite, as shown in Sections 7.7 and 7.8.

### 7.9.1 Perfect Signal Detection

This paradox may arise in communication theory when one is testing between the hypothesis that the received signal R(t) = S(t) + N(t) (t in T) and the hypothesis that R(t) = N(t) (t in T), where N(t) (t in T) is random noise. If $P_{N_T} \perp P_{S_T+N_T}$ it is then possible to detect with arbitrarily small error, whether there is a signal S present, by observing the received signal over an arbitrarily small interval of length T.

# VIII. RELATIONSHIPS BETWEEN CERTAIN RANDOM PROCESSES AND THE SINGULARITY BETWEEN PROBABILITY MEASURES

In this chapter several types of random processes (stochastic processes) over intervals are investigated. Ergodic processes with analytic covariance functions, singular and completely deterministic processes, and gaussian processes with proportional covariance functions are discussed and related to singularity between probability measures. These processes are also related to errorless discrimination, infinite expected mutual information, infinite expected weight of evidence, and infinite divergence, over arbitrarily small intervals of time.

## 8.1 Definitions

Let $X(t)$ $(-\infty < t < \infty)$ represent a real random process for the following definitions which describe the stationarity properties and the first and second moments of the process.

### 8.1.1 Strictly Stationary Random Process

A strictly stationary random process $X(t)$ $(-\infty < t < \infty)$ is a random process for which the joint distribution of the set $\{X(t_i)\}$ $(i = 1, \ldots, k)$ is the same as the joint distribution of $\{X(t_i + h)\}$ $(i = 1, \ldots, k)$ for all sets $\{t_i\}$ $(i = 1, \ldots, k)$ and for all positive integers k and constants h. It follows that the moments of $X(t)$ at any point t do not depend on t since they are the same at $X(t + h)$.

### 8.1.2 Mean Function

Define $m(t) = E[X(t)]$ $(-\infty < t < \infty)$.

### 8.1.3 Covariance Function

The **covariance function** (or the autocovariance function) is defined as

$$\gamma_X(t,s) = E\{[X(t) - m(t)][X(s) - m(s)]\} . \qquad (8.1.3a)$$

When X is taken for granted the subscript X is omitted. If $Y(s)$ $(-\infty < s < \infty)$ is another random process then the cross-covariance function is defined by

$$\gamma_{XY}(t,s) = E\{[X(t) - m_X(t)][Y(s) - m_Y(s)]\} . \qquad (8.1.3b)$$

### 8.1.4 Variance Function

The variance function is defined by

$$\sigma^2(t) = \gamma(t,t) = E[X(t) - m(t)]^2 . \qquad (8.1.4)$$

A gaussian process is completely defined by its mean and covariance function. Therefore, at any point, it is *a fortiori* completely defined by its mean and variance.

### 8.1.5 Correlation Function

The correlation function (also called the autocorrelation function) is defined by

$$\rho(t,s) = \frac{\gamma(t,s)}{\sigma(t)\sigma(s)} . \qquad (8.1.5)$$

### 8.1.6 Covariance-Stationary Random Process

A covariance-stationary random process (also called stationary in the wide sense, weakly stationary, or second-order stationary) has the property that $\gamma(t,s)$ is a function of $t-s$ alone, denoted by $\gamma(t-s)$. In particular, the variance function $\sigma^2 = \gamma(t,t) = \gamma(0)$ does not depend on t. (The mean function $m(t)$ may still depend on t.) The correlation function can therefore be written

$$\rho(\tau) = \gamma(\tau)/\sigma^2 . \qquad (8.1.6)$$

Since X is real, $\gamma(\tau)$ and $\rho(\tau)$ must be even functions of $\tau$.

A strictly stationary random process is covariance-stationary since the joint distribution of $X(t)$ and $X(t + \tau)$ is the same for all t and therefore depends only on $\tau$.

A gaussian random process is strictly stationary if and only if it is covariance-stationary and in addition its mean function is mathematically independent of t.

## 8.2 Spectral Analysis

For each covariance-stationary process, having a continuous covariance function, there exists a real nondecreasing nonrandom function of a real variable $\omega$ denoted by $F^*(\omega)$, called the spectral distribution function, such that $F(-\infty) = 0$, and

$$\gamma(\tau) = \int_{-\infty}^{\infty} e^{i2\pi\tau\omega} dF^*(\omega) = 2\int_0^{\infty} \cos 2\pi\tau\omega \, dF^*(\omega), \tag{8.2.1}$$

where $-\infty < \tau < \infty$ and where $\omega$ is the frequency measured in cycles per second if $\tau$ is measured in seconds. Let $\lambda = 2\pi\omega$, so that $\lambda$ is measured in radians per second and

$$\gamma(\tau) = \frac{1}{2\pi} \int_{-\infty}^{\infty} e^{i\lambda\tau} dF^*(\lambda) \; . \tag{8.2.2}$$

It may be observed that

$$\sigma^2 = \gamma(0) = \int_{-\infty}^{\infty} dF^*(\omega) \; . \tag{8.2.3}$$

If $F^*(\omega)$ is a step function then in (8.2.1) $\gamma(\tau)$ can be expressed as an infinite sum of periodic components having positive mass $p^*(\omega_j)$ at the frequencies $\omega_j$. That is,

$$\gamma(\tau) = \sum_{j=1}^{\infty} \cos (2\pi\tau\omega_j) p^*(\omega_j) . \tag{8.2.4}$$

If $F^*(\omega)$ is absolutely continuous then its non-negative derivative exists, almost everywhere, and will be denoted by $P_X^*(\omega)$ when $X(t)$ is the underlying random process. $P_X^*(\omega)$ is called the <u>spectral density function</u>. Some authors call $P_X^*(\omega)$ the spectrum and reserve the name spectral density function for

$$f_X^*(\omega) = \frac{P_X^*(\omega)}{\sigma^2} \tag{8.2.5}$$

where $f_X^*(\omega) \geq 0$ and $\int_{-\infty}^{\infty} f_X^*(\omega)d\omega = 1$. When $F^*(\omega)$ is absolutely continuous, $\gamma_X(\tau)$ can be expressed as

$$\gamma_X(\tau) = \int_{-\infty}^{\infty} e^{i2\pi\tau\omega} P_X^*(\omega)d\omega \tag{8.2.6}$$

where

$$P_X^*(\omega) = \int_{-\infty}^{\infty} e^{-i2\pi\omega\tau}\gamma_X(\tau)d\tau = \int_{-\infty}^{\infty} e^{-i\lambda\tau}\gamma_X(\tau)d\tau . \tag{8.2.7}$$

Therefore, $\lim_{|\tau|\to\infty} \gamma_X(\tau) = 0$, when $P_X^*(\omega)$ exists. A sufficient condition for $F^*(\omega)$ to be absolutely continuous is that $\int_{-\infty}^{\infty} |\gamma(\tau)|d\tau$ is finite. The two relationships (8.2.6) and (8.2.7) between $\gamma(\tau)$ and $P_S^*(\omega)$ are called the Wiener-Khintchine relations in honor of the pioneering work of Wiener (1930) and Khintchine (1934). The variance may now be expressed as in (8.2.3) by

$$\sigma_X^2 = \int_{-\infty}^{\infty} P_X^*(\omega)d\omega \tag{8.2.8}$$

and thus the variance, often called the mean power, is decomposed into the integral of the spectral density function, a kind of "analysis of variance".

$F^*$ has all the properties of an ordinary distribution function (except that $F^*(\infty)$ need not be equal to 1) so that, as in Section 3.2.2, it can be decomposed as follows

$$F^*(\omega) = F_{ac}^*(\omega) + F_{sc}^*(\omega) + F_d^*(\omega)$$

where $F_{ac}^*(\omega)$ is the absolutely continuous part $F_{sc}^*(\omega)$ is the singularly continuous part, and $F_d^*(\omega)$ is the discrete part. [Doob (1953, p. 488).] Therefore $\gamma(\tau)$ can also be expressed as

$$\gamma(\tau) = \gamma_1(\tau) + \gamma_2(\tau) + \gamma_3(\tau)$$

where $\gamma_1$, $\gamma_2$, and $\gamma_3$ correspond to $F^*_{ac}$, $F^*_{sc}$, and $F^*_d$ respectively. $\gamma_1(\tau)$ is expressed by (8.2.6) and has the property that $\gamma_1(\tau) \to 0$ as $\tau \to \infty$, $\gamma_3(\tau)$ is expressed by (8.2.4), whereas $\gamma_2(\tau)$ probably has no physical interpretation.

For frequencies $\omega_1$ and $\omega_2$ where $\omega_1 < \omega_2$, $F^*(\omega_2) - F^*(\omega_1)$ may be interpreted as a measure of the contribution of the frequency band $\omega_1$ to $\omega_2$ to the random process.

Suppose two random processes X(t) and Y(t) are related by the equation

$$Y(t) = \int_{-\infty}^{\infty} g(s)X(t - s)\,ds \tag{8.2.9}$$

where g(s) is some non-random function such that $g(s) = 0$ if $s < 0$. Let $Y^*(\omega)$, $g^*(\omega)$, and $X^*(\omega)$ be the fourier transforms of Y(t), g(t), and X(t). For instance,

$$Y^*(\omega) = \int_{-\infty}^{\infty} e^{-i2\pi\omega t}Y(t)\,dt \ .$$

Then

$$Y^*(\omega) = g^*(\omega)X^*(\omega) \ .$$

If X(t) and Y(t) are covariance-stationary processes then

$$P^*_Y(\omega) = |g^*(\omega)|^2 P^*_X(\omega) \tag{8.2.10}$$

where $P^*_Y(\omega)$ and $P^*_X(\omega)$ are the spectral density functions of Y(t) and X(t). The function g(t) is called the <u>impulse response function</u> and $g^*(\omega)$ is called the <u>frequency response function</u> or <u>transfer function</u>.

<u>Exercise.</u> (i) If two statistically independent processes, having spectral densities, are multiplied together (perhaps a random carrier wave and a gaussian signal), the spectral density of the product is the convolution of the individual spectral densities. (ii) If X(t) is the average of a large number of statistically independent identically distributed processes, then $e^{X(t)}$ has bell-shaped covariance and spectral density functions. (A "law of large

numbers" for random processes which we have not seen in the literature and which could presumably be generalized.)

Two references for this Section 8.2 are Parzen (1962, pp. 107, 110) and Jenkins and Watts (1968, pp. 217, 218).

### 8.2.1 Rational Spectral Density Functions and their Relationships to Linear Systems and Markov Processes

Several of the results in this monograph require that certain spectral density functions are rational functions of $\omega$. Whenever this is true they must be rational functions of $\omega^2$. [See, for instance, Helstrom (1968b, p. 57).]

Wiener (1949, p. 37) showed that if the covariance function $\gamma(\tau)$ $(-\infty < \tau < \infty)$ is such that

$$\int_{-\infty}^{\infty} \gamma^2(\tau)\,d\tau < \infty$$

then $\gamma(\tau)$ can be represented in terms of a countable linear combination of Laguerre functions which have rational fourier transforms.

Granger and Hatanaka (1964, pp. 37, 38), for example, point out that any real process having a continuous spectrum may be approximated as closely as we wish by a process having a rational spectral density function.

Davenport and Root (1958, p. 105) and Martel and Mathews (1961) have shown that the spectral density of the output from a lumped-parameter linear filter (or system) in which the input is white noise is always a rational function of $\omega$. (White noise will be defined in Section 8.5.)

By a <u>linear filter</u> or <u>linear system</u> is meant a system such that, if the input to the system is $af(t) + bg(t)$, then the output is $ar(t) + bs(t)$, where $r(t)$ and $s(t)$ are the outputs corresponding to the inputs $f(t)$ and $g(t)$ respectively. [Lathi (1965, pp. 2, 3).]

A <u>lumped-parameter system</u> is a circuit with the energy stored in isolated blocks of capacitance, resistance, inductance, etc. The behavior of such systems is described by ordinary differential equations. A distributed system is one which is not lumped, such as a transmission line. The behavior of a distributed system is described by partial differential equations. At high frequencies lumped-parameter systems in practice become distributed systems.

If the linear system is time invariant, so that the output at time $t_2$ depends only $t_2 - t_1$ where $t_1$ is the time of the system's input, then the output $Y(t)$ can always be represented by a convolution as in Equation (8.2.9) where $X(t)$ is the input to the system. The output of such a system is always gaussian if the input is gaussian. See Jenkins and Watts (1968, p. 34) and Breiman (1969, pp. 274-276). For example, if a linear system is defined by a first-order differential equation

$$\frac{dY(t)}{dt} + \alpha Y(t) = X(t) \qquad (8.2.1.1)$$

where $\alpha$ is a positive constant, then

$$Y(t) = \int_{\infty}^{t} e^{-\alpha(t - s)} X(s)\,ds \qquad (8.2.1.2)$$

so that $g(t) = e^{-\alpha t}$ $(0 < t < \infty)$ and $g(t) = 0$ when $t < 0$ in Equation (8.2.9). Breiman (1969, p. 286) shows that a stationary gaussian process $Y(t)$ is a Markov process if and only if it is the output of a first-order linear system defined by (8.2.1.1) where the input is white noise with constant spectral density function (mathematically independent of $\omega$). White noise is defined and discussed in Section 8.5. A Markov process is one where the future values given the present and the past values, depend only on the present values. For the linear system just defined

$$|g^*(\omega)|^2 = \left|\int_0^{\infty} e^{-i2\pi\omega t} e^{-\alpha t} dt\right|^2 = \left|\frac{1}{\alpha + i2\pi\omega}\right|^2 = \frac{1}{\alpha^2 + (2\pi\omega)^2} \qquad (8.2.1.3)$$

and from Equation (8.2.10)

$$P_Y^*(\omega) = P_X^*(\omega)\,|g^*(\omega)|^2 = \frac{k}{\alpha^2+(2\pi\omega)^2} \qquad (8.2.1.4)$$

which is a <u>rational</u> function of $\omega^2$. The corresponding covariance function of Y is therefore

$$\gamma_Y(\tau) = \int_{-\infty}^{\infty} e^{i2\pi\omega\tau} P_Y^*(\omega)\,d\omega = \sigma_Y^2 e^{-\alpha|\tau|} \; . \qquad (8.2.1.5)$$

It can be concluded that a stationary gaussian process $Y(t)$ $(-\infty < t < \infty)$ is a Markov process if and only if $P_Y^*(\omega)$ and $\gamma_Y(\tau)$ are of the forms in (8.2.1.4) and (8.2.1.5), respectively.

Any gaussian stationary process with a rational spectral density function can always be represented by a component of a multidimensional stationary gaussian Markov process of dimension equal to one half the degree of the denominator of the rational function [Helstrom (1968, p. 57) and Pinsker (1964, pp. 171-172)]. A multidimensional process may be Markov, even though its components are not. [Breiman (1969, p. 297).] A multidimensional stationary gaussian process, $\mathbf{Y}'(t) = (Y_1(t), \ldots, Y_n(t))$, is a Markov process if and only if it is the output of the Langevin linear system defined below, where the input process $\mathbf{X}(t)' = (X_1(t), \ldots, X_n(t))$ has the property that each of its components has a zero mean and the cross spectral density matrix of its components is a constant. (White noise.) The Langevin linear system is defined as in Equation (8.2.1.1),

$$\frac{d\mathbf{Y}(t)}{dt} + \mathbf{\Lambda Y}(t) = \mathbf{X}(t) \; ,$$

where $\Lambda$ is a nonsingular $n \times n$ matrix of constants. Then we have the following set of integral equations.

$$\mathbf{Y}(t) = \int_{-\infty}^{t} e^{-\mathbf{\Lambda}(t-s)} \mathbf{X}(s)\,ds$$

where

$$e^{-\Lambda t} = 1 - \Lambda t + \frac{1}{2!}\Lambda^2 t^2 - + \ldots$$

is the unique solution to the Markov matrix M(t) defined by

$$M(t+s) = M(s)M(t),$$

when

$$e^{-\Lambda} = M(1).$$

The covariance matrix $\gamma(\tau)$ of the components of Y(t) is then

$$\gamma(\tau) = \gamma(0)e^{-\Lambda'|\tau|}$$

where $\Lambda'$ is the transpose of $\Lambda$. This is expressed similarly to Equation (8.2.1.5). Each component of the spectral density matrix of Y(t) is a rational function of $\omega$. For a reference to this multidimensional case see, for instance, Helstrom (1968).

## 8.3 Ergodic Process

Let X(t) $(-\infty < t < \infty)$ be a strictly stationary random process and let $Y(t) = g(X(t))$ be any function of X(t) such that the constant mean $m = E[Y(t)]$ exists. This process is ergodic if with probability 1 any sample function y(t) $(-\infty < t < \infty)$ satisfies

$$\lim_{T\to\infty} \frac{1}{T}\int_0^T y(t)\,dt = m .$$

As an example, let $Y(t) = X(t)X(t+\tau)$ where $E[X(t)] = 0$, then $m = \gamma(\tau)$ $= E[X(t)X(t+\tau)]$. For a further discussion see Doob (1953, Chap. 11 Arts .1 and 8), Grenander (1950, Art. 5.10) and Davenport and Root (1958, pp. 67, 68).

Suppose X(t) $(-\infty < t < \infty)$ is a random process. Let a, b, and d, where $a \le b$, be real constants. Define the sets of random variables,

$$U = \{X(t)\} \quad (a \le t \le b)$$

and

$$V = \{X(t)\} \quad (a + d \le t \le b + d).$$

Suppose that

$$\lim_{d\to\infty} \frac{P[U \text{ in } A, V \text{ in } B]}{P[U \text{ in } A]P[V \text{ in } B]} = 1$$

for all values of a, b, and d, where $a \le b$, when A and B are arbitrary measurable sets. Then it is said that the sets of random variables on two widely separated intervals are almost independent and the process is said to have the asymptotic independence property over intervals. Breiman (1969, p. 296) calls a process having this property a non-deterministic process. This property is also mentioned by Kolmogorov (1963, p. 311) and (1956, p. 107).

When, with probability 1, the future is completely determined by the past, we shall call the process completely deterministic.

These two types of processes will be discussed further in Section 8.8.

Breiman (1969, p. 317) mentions that any strictly stationary non-deterministic process is ergodic. A strictly stationary ergodic process may be completely deterministic however, as we shall see in Section 8.7. It is also true that a non-deterministic process must have a continuous spectral distribution function so that it does not contain any periodic components.

Grenander (1950, p. 257) shows that a strictly stationary gaussian process is ergodic if and only if it has an absolutely continuous spectral distribution function. If such a process is completely deterministic it still must have the almost independence property when one point is widely separated from any given finite set of points, as we shall see in the following theorem.

Theorem 8.3 If $X(t)$ $(-\infty < t < \infty)$ is a covariance stationary ergodic gaussian process with zero mean, then the value of the process at a point widely separated from any finite set of points is nearly independent of the values at these points. (It is not essential that the mean be zero, but it simplifies the notation.)

Proof: Grenander (1950, p. 257) shows such a process must have a finite spectral density function

$$P_X^*(\omega) = \int_{-\infty}^{\infty} e^{-i2\pi\omega\tau}\gamma(\tau)d\tau$$

which implies that $\gamma(\tau) \to 0$ as $|\tau| \to \infty$. Let

$$\mathbf{X}_n' = (X(t_1), \ldots, X(t_n))$$

represent the random vector of the process at the n points $t_1, \ldots, t_n$ and let $X(s)$ represent the random value of the process at any point s. It now follows that

$$\lim_{s\to\infty} E[\mathbf{X}_n X(s)] = \mathbf{0}_n$$

where $\mathbf{0}_n$ is an n-dimensional column vector of zeros. If Y is any random variable independent of $\mathbf{X}_n$ then $E[\mathbf{X}_n Y] = \mathbf{0}_n$. Since any multivariate normal distribution with zero mean is completely determined by its covariance matrix, it follows that $\mathbf{X}_n$ and $X(s)$ are asymptotically independent.

## 8.4 Band-Limited White Noise

If $P_X^*(\omega) = 0$ when $|\omega| > W$, then $X(t)$ is said to be band-limited by the frequency W. A covariance-stationary random process $N_W(t)$ with mean zero is called band-limited white noise if it has a spectral density function $P_{N_W}^*(\omega)$ such that $P_{N_W}^*(\omega) = K$ for all $|\omega| \leq W$ for some W, and $P_N^*(\omega) = 0$ when $|\omega| > W$. In this case,

$$\gamma(\tau) = K\int_{-W}^{W} e^{i2\pi\omega\tau}d\omega = 2K\int_0^W \cos(2\pi\omega\tau)d\omega = 2K\frac{\sin(2\pi\omega\tau)}{2\pi\tau} = \sigma_N^2\,\frac{\sin(2\pi W\tau)}{2\pi\tau W} = \sigma_N^2\psi(\tau), \quad (8.4.1)$$

where

$$\psi(\tau) = \frac{\sin(2\pi W\tau)}{2\pi\tau W} ,$$

and

$$\sigma_N^2 = \gamma(0) = \int_{-W}^{W} P^*_{N_W}(\omega)d\omega = 2WK .$$

It may be observed that $\gamma(\tau) = 0$ when $\tau = \frac{j}{2W}$ $(j = 0, \pm1, \pm2, \ldots)$. Therefore the values of $N(t)$ are uncorrelated at a set of points at distances apart of $\frac{1}{2W}$. The fraction $\frac{1}{2W}$ is half the time required to complete one cycle at the frequency $W$. $N(t)$ $(-\infty < t < \infty)$ can be completely determined, for this particular sample function, by its values at the set of uncorrelated points at distance apart of $\frac{1}{2W}$ by the following expression.

$$N_W(t) = \sum_{n=-\infty}^{\infty} N_W\left(\frac{n}{2W}\right) \frac{\sin \pi(2Wt-n)}{\pi(2Wt-n)} . \tag{8.4.2}$$

This expression, for band-limited signals in general, was perhaps first proved by Whittaker (1915) and introduced into communication theory by Nyquist (1928). See Belyaev (1959, p. 406) for another proof. The ideas of this section can be found, for example, in Good and Doog (1958).

## 8.5 White Noise and the Dirac Delta Function

A covariance-stationary random process $N(t)$ with mean zero is called white noise if it has a uniform (constant) continuous spectral density function $P_N^*(\omega) = K$ $(-\infty < \omega < \infty)$. This is a generalized random process with $\gamma(\tau) = K\delta(\tau)$, where $\delta(0) = \infty$ and $\delta(\tau) = 0$ if $\tau \neq 0$.

$$\int_{-\varepsilon}^{\varepsilon} \delta(\tau)d\tau = \int_{-\infty}^{\infty} \delta(\tau)d\tau = 1$$

for all $\varepsilon > 0$. For any function $g(t)$

$$\int_{-\infty}^{\infty} g(t)\delta(t-t_o)dt = g(t_o) \tag{8.5.0}$$

and

$$\int_{-\infty}^{\infty} g(t)\delta(t)dt = g(0) .$$

White noise is a completely random or "uncorrelated" process in the sense that

$$\rho(\tau) = \begin{cases} \dfrac{\gamma(\tau)}{\gamma(0)} = \dfrac{K\delta(\tau)}{K\delta(0)} = \dfrac{0}{\infty} = 0 & (\text{if } \tau \neq 0) \\ \dfrac{\gamma(0)}{\gamma(0)} \text{ which is defined as } 1 & (\text{if } \tau = 0). \end{cases} \tag{8.5.1}$$

The spectral density function $P_N^*(\omega)$ is from (8.2.7),

$$P_N^*(\omega) = \int_{-\infty}^{\infty} \gamma(\tau)e^{-i2\pi\omega\tau}d\tau = Ke^{-i2\pi\omega 0} = K \tag{8.5.2}$$

where $(-\infty < \omega < \infty)$. White noise has no meaning by itself but by using the techniques of generalized functionals [see Gelfand and Yaglom (1959)] the functional $N(\phi)$ can be defined by

$$N(\phi) = \lim_{W\to\infty} \int_{-\infty}^{\infty} N_W(t)\phi(t)dt = \int_{-\infty}^{\infty} N(t)\phi(t)dt \tag{8.5.3}$$

where $N_W(t)$ is band-limited white noise, and $\phi$ ranges over some space $\Phi$ of functions such as the space of all infinitely differentiable functions. $N(\phi)$ is a random variable with mean

$$E[N(\phi)] = \int_{-\infty}^{\infty} E[N(t)]\phi(t)dt = 0 \tag{8.5.4}$$

and with variance

$$E[N^2(\phi)] = \int_{-\infty}^{\infty}\int_{-\infty}^{\infty} E[N(s)N(t)]\phi(s)\phi(t)ds\ dt$$

$$= K\int_{-\infty}^{\infty} \phi(s)\ (\int_{-\infty}^{\infty} \phi(t)\delta(t-s)dt)ds = K\int_{-\infty}^{\infty} \phi^2(s)ds. \tag{8.5.5}$$

## 8.6 Periodic Band-Limited White Noise

The results of this section are from Good and Doog (1958).

Let $N_W(t)$ be a covariance-stationary random process with mean zero that is periodic with period T. Any periodic function with period T, such as $\gamma(\tau)$, can always be expressed as a fourier series as follows,

$$\gamma(\tau) = \sum_{n=-\infty}^{\infty} e^{i2\pi\tau(n/T)} p^*(\tfrac{n}{T}). \qquad (8.6.1)$$

See, for example, Jenkins and Watts (1968, p. 26). Therefore $F^*(\omega)$ in (8.2.1) is discrete and there is positive mass at the multiples of the fundamental frequency (1/T).

If the discrete spectral mass function, $p^*(\frac{n}{T})$, is such that $p^*(\frac{n}{T}) = K$, $|\frac{n}{T}| \le W$, for some W, and $p^*(\frac{n}{T}) = 0$ when $\frac{n}{T} > W$, then $N_W(t)$ is called periodic band-limited white noise. Suppose that $\frac{n_o-1}{T} \le W < \frac{n_o}{T}$, where $n_o$ is a positive integer. Then from (8.6.1),

$$\gamma(\tau) = K \sum_{n=-(n_o-1)}^{n_o-1} e^{i2\pi(\frac{n}{T})\tau} = K(1 + 2\sum_{n=1}^{n_o-1} \cos 2\pi(\tfrac{n}{T})\tau)$$

$$= \frac{K \sin [\pi(2n_o-1)\tau/T]}{\sin (\pi\tau/T)}, \qquad (8.6.2)$$

by Lagrange's identity. [See Brillouin (1956, p. 95).] Since

$$\sigma_N^2 = \gamma(0) = K[1 + 2(n_o-1)] = K(2n_o-1)$$

it follows that

$$K = \sigma_N^2/(2n_o-1). \qquad (8.6.3)$$

It may be observed that $\gamma(\tau)$ is a continuous function of $\tau$ even though the spectrum is discrete. $\gamma(\tau) = 0$ when $\tau - jT/(2n_o-1)$ where $(j = 0, \pm 1, \pm 2, \pm \ldots)$. The values of N(t) are uncorrelated, therefore, at a set of points at distances apart of $T/(2n_o-1)$. $N_W(t)$ can be completely determined by any set of $(2n_o-1)$ points in the interval [0,T] for this particular sample function. In particular, at the uncorrelated points,

$$N(t) = \sum_{j=0}^{2n_o-2} N\left(\frac{jT}{2n_o-1}\right) \frac{\sin \pi\left(j - \frac{2n_o-1}{T}\right)}{(2n_o-1)\sin \pi\left(\frac{j}{2n_o-1} - \frac{t}{T}\right)} \tag{8.6.4}$$

when $0 \leq t \leq T$.

## 8.7 Ergodic Strictly Stationary Random Processes with Analytic Covariance Functions

If the covariance function $\gamma_1(\tau)$ $(-\infty < \tau < \infty)$ of a strictly stationary random process $X_1(t)$ $(-\infty < t < \infty)$ is an analytic function of $\tau$, then from Belyaev (1959), with probability one, any sample function is an analytic function of t. (An analytic function has the property that all its derivatives exist and it can be expanded in a Taylor series about any point.) Therefore, by using the values of almost any sample function over any arbitrarily small interval, the entire sample function can be constructed for $-\infty < t < \infty$ by using the Taylor expansion about any point in T.

If the process is ergodic then, with probability one, all the moments can be determined from this sample function, so that at least for gaussian processes it is then possible to discriminate with zero error between $X_1(t)$ and any other random process $X_2(t)$ by using the values over any arbitrarily small interval T. [Belyaev (1959, p. 409).] This process is ergodic even though it is completely deterministic, as mentioned in Section 8.3. Therefore, $P_{1T} \perp P_{2T}$, from Section 7.9, so that $W_T(H_1/H_2) = \infty$ from Section 7.4, and $J_T(1,2) = \infty$ from Section 7.5.

### 8.7.1 Strictly Stationary Band-Limited Gaussian Processes with Absolutely Continuous Spectral Distribution Functions

The covariance function of a covariance-stationary band-limited random process is analytic since

$$\gamma(\tau) = \int_{-W}^{W} e^{i2\pi\tau\omega} dF^*(\omega)$$

from (8.2.1) and all the derivatives with respect to $\tau$ exist. A strictly stationary gaussian process with an absolutely continuous spectral distribution function is ergodic, as mentioned in Section 8.3. Therefore, this is a special case of the process discussed in Section 8.7.

### 8.7.1.1 Gaussian Band-Limited White Noise

Gaussian band-limited white noise is a strictly stationary band-limited gaussian process with an absolutely continuous spectral distribution function, so that this is an example of the process in Section 8.7.1.

Periodic gaussian band-limited white noise is not ergodic since the spectral distribution function is not absolutely continuous, but discrete, as pointed out in Section 8.6. Therefore, the statistical moments of the process cannot be determined from a single sample function, as in Section 8.7, so that, from Section 7.8, $P_{1T} \parallel P_{2T}$, $W_T(H_1/H_2) < \infty$, and $J_T(1,2) < \infty$, and this is not an example of Section 8.7.

### 8.7.1.2 Perfect Signal Detection

If $R_1(t) = S(t) + N(t)$ and $R_2(t) = N(t)$ (t in T), and N is gaussian band-limited white noise, then $P_{N_T} \perp P_{S_T+N_T}$ and $W_T(H_N/H_{S+N}) = \infty$, and $J_T(N, S+N) = \infty$ regardless of whether S(t) is a random process. This result is a special case of the processes described in Section 8.7 and the first part of Section 8.7.1.1. $P_{N_T}$ and $P_{S_T+N_T}$ represent the probability measures for N(t) and S(t) + N(t), t in T. As discussed in Section 7.9.1 it is then possible to detect whether there is a signal S present, with zero error, by observing the process over an arbitrarily small interval T.

## 8.8 Prediction

Let X(t) $(-\infty < t < \infty)$ be a strictly stationary random process with a constant variance $\sigma^2$ and let r and s denote any two points of the process such that $r < s$. The minimum mean-squared-error predictor $\hat{X}(s - r)$ of $X(s)$ for the set of random variables $\{X(t)\}$ $(t \le r)$, and the corresponding prediction error $e(s - r)$ are

$$\hat{X}(s - r) = E[X(s) | \{X(t)\}(t \le r)] \tag{8.8.0.1}$$

and

$$e(s - r) = E[X(s) - \hat{X}(s - r)]^2 = E\{V[X(s) | \{X(t)\}(t \le r)]\}, \tag{8.8.0.2}$$

where V denotes the variance operator.

It can be shown [Parzen (1962b, p. 55)] that

$$\sigma^2 = V[X(s)] = e(s - r) + V[\hat{X}(s - r)] . \tag{8.8.0.3}$$

Therefore, $e(\ell) \le \sigma^2$ for all lag times $\ell$. Moreover $e(\ell)$ is a non-decreasing function of $\ell$ and either $e(\ell) = 0$ for all $\ell > 0$ or else $e(\ell) > 0$ for all $\ell > 0$. Let

$$e(\infty) = \lim_{\ell \to \infty} e(\ell) ;$$

then

$$\sigma^2 \ge e(\infty) \ge e(\ell) \ge e(0) = 0$$

for all $\ell > 0$.

If $e(\ell) > 0$ for all $\ell > 0$, then there are two possibilities, either $e(\infty) = \sigma^2$ or $e(\infty) < \sigma^2$. If the process is a non-deterministic process (which was defined in Section 8.3) then with probability 1

$$\lim_{s \to \infty} V[X(s) | \{X(t) = x(t)\}(t \le r)] = \sigma^2$$

independent of any sample set $\{x(t)\}$ $(t \le r)$ because of the asymptotic independence property defined in Section 8.3. Therefore $e(\infty) = \sigma^2$. A non-deterministic process must be erogidc as mentioned in Section 8.3. Breiman calls the process partially deterministic if $0 < e(\infty) < \sigma^2$.

If $e(\ell) = 0$ for all $\ell \geq 0$, then it follows that for any two points, r and s, of the process, where $r \leq s$, we have $X(s) = \hat{X}(r - s)$ with probability 1. That is, for almost all sample functions, the past completely determines the future. Such a process is called completely deterministic, as mentioned in Section 8.3. Any analytic process, such as a covariance-stationary band-limited process is completely deterministic. Completely deterministic ergodic processes result in perpendicular probability measures as shown in Section 8.7.

Most of the ideas of this section can be found in Breiman (1969, pp. 295, 296).

### 8.8.1 Linear Prediction

As in much of this monograph we shall here summarize some salient points from the literature, listed at the end of the section, without detailed discussion.

Let $X(t)$ $(-\infty < t < \infty)$ be any covariance-stationary process with a covariance function $\gamma_X(\tau)$ $(-\infty < \tau < \infty)$, a mean zero, and an absolutely continuous spectral distribution function $P_X^*(\omega)$ $(-\infty < \omega < \infty)$. Throughout Section 8.8.1 we shall assume the process satisfies these assumptions just mentioned. We shall define a <u>linear predictor</u> for the set of random variables $\{X(t)\}$ $(t \leq r)$ as the following linear operator,

$$\int_{-\infty}^{r} g(t)dX(t) \;, \qquad (8.8.1.1)$$

where $g(\cdot)$ is any function such that $\int_{-\infty}^{\infty} g^2(t)dt < \infty$. This integral includes linear combinations of $X(t_i)$ $(t_i \leq r)$ as well as $\int_{-\infty}^{r} g(t)X(t)dt$.

Let r and s be any two points of the process such that $r < s$. The minimum mean-squared-error <u>linear predictor</u> $\hat{X}_L(s - r)$ of $X(s)$ for the set of random variables $\{X(t)\}$ $(t \leq r)$ is of the form

$$\hat{X}_L(s - r) = \int_{-\infty}^{r} R(r - t)dX(t) \qquad (8.8.1.2)$$

where the function $R(\cdot)$ depends on the lag time $s - r$. We may think of $\hat{X}_L(s - r)$ as the output of a time-invariant linear system whose impulse response function is $R(r - t)$ with input $X(t)$, as with equation (8.2.9) and described in Section 8.2.1. $R(\cdot)$ may be found by solving the Wiener-Hopf equation,

$$\gamma_X(\varepsilon + s - r) = \int_0^\infty R(v)\gamma_X(\varepsilon - v)\,dv \qquad (8.8.1.3)$$

where $\varepsilon$ is any non-negative constant. A similar formula holds for linear combinations of $X(t_i)$ $(t_i \le r)$. The corresponding linear prediction error is

$$e_L(s - r) = E[X(s) - \hat{X}_L(s - r)]^2 \; . \qquad (8.8.1.4)$$

If $X(t)$ $(-\infty < t < \infty)$ defined above is a gaussian process then $\hat{X}_L(\ell) = \hat{X}(\ell)$, and $e_L(\ell) = e(\ell)$ for all $\ell \ge 0$, so all the remarks concerning $e(\ell)$, for the general prediction problem, apply here also.

If the process is strictly stationary, but not gaussian, then the optimal predictor is not usually a linear predictor and therefore $\hat{X}_L(\ell)$ and $e_L(\ell)$ are not usually equal to $\hat{X}(\ell)$ and $e(\ell)$ respectively, but $e_L(\ell) > e(\ell)$ for all $\ell > 0$. If the process is non-deterministic then $e_L(\infty)$ is not equal to $\sigma^2$ unless $\hat{X}_L(\ell) = \hat{X}(\ell)$ for all $\ell > 0$. If $\hat{X}_L(\ell) \ne \hat{X}(\ell)$ then

$$\sigma^2 \ne e_L(s - r) + V[\hat{X}_L(s - r)] \quad ,$$

and the deductions from (8.8.0.3) cannot be made, but it is still true that $e_L(\ell)$ is a non-decreasing function of $\ell$ and either $e(\ell) = 0$ for all $\ell > 0$ or else $e(\ell) > 0$ for all $\ell > 0$.

If $e_L(\ell) > 0$ for all $\ell > 0$ then the process is said to be regular. If a process is regular then with probability 1, the future cannot be completely determined by a linear operation on the past in the form of the expression (8.8.1.1). It can still be a completely deterministic process, however. In fact, for most non-gaussian completely deterministic strictly stationary processes satisfying the assumptions of this section, $\hat{X}(\ell) \neq \hat{X}_L(\ell)$, in which case they are regular because $e_L(\ell) > e(\ell) = 0$. All non-deterministic or partially deterministic processes satisfying the assumptions of this section must be regular because $e_L(\ell) \geq e(\ell) > 0$ for all $\ell > 0$. Vainshtein and Zubakov (1962, p. 70) discuss regular processes from another point of view.

Any process satisfying the assumptions of this section is regular if and only if $P_X^*(\omega) > 0$ almost everywhere with respect to Lebesgue measure, and the Lebesgue integral,

$$\int_{-\infty}^{\infty} \frac{|\log P_X^*(\omega)|}{1 + \omega^2} d\omega < \infty . \tag{8.8.1.5}$$

This is usually called the Paley-Wiener criterion.

The regular processes include the processes with rational spectral density functions (Vainshtein and Zubakov, 1962).

Vainshtein and Zubakov (1962, pp. 368, 369) give two examples which illustrate that it is possible to have perfect detection of arbitrarily weak non-random signals over an infinite range of t, even when the noise is regular.

If $e_L(\ell) = 0$ for all $\ell \geq 0$ the process is called singular. Then for almost all sample functions, the future can be completely determined by a linear operation on the past in the form of the expression (8.8.1.1). All singular processes that are strictly stationary are completely deterministic because $e_L(\ell) \geq e(\ell)$ for all $\ell \geq 0$ so that $e(\ell) = 0$. As mentioned previously, most non-gaussian completely deterministic processes are not singular, however. Strictly stationary singular ergodic processes result in perpendicular probability measures.

All processes with a spectral distribution function that is discrete, so that $\gamma(\tau)$ can be expressed as a sum of periodic components, are singular, and therefore completely deterministic. [See, for instance, Breiman (1962, p. 306).]

Any process satisfying the assumptions of this section is singular if and only if either $P_X^*(\omega) = 0$ over some interval or the Paley-Wiener integral (8.8.1.5) is infinite. [Breiman (1969, pp. 306, 307).]

Vainshtein and Zubakov (1962, pp. 20, 21, 69, 70, and 365) give some examples of processes that are not band limited or analytic, and where $P_X^*(\omega) > 0$ everywhere, but are still singular because $P_X^*(\omega)$ falls off sufficiently fast for large frequencies, resulting in infinite values for the Paley-Wiener integral.

Selin (1962, p. 326) illustrates two examples of singular processes that are not band-limited and are limiting forms of regular processes with rational spectral density functions.

Some references for Section 8.8.1, some of which also discuss the discrete-parameter processes, are Breiman (1969, pp. 300-306), Kolmogorov (1941a, 1941b), Wiener (1949, pp. 37, 55), Doob (1952, pp. 583, 584), Vainshtein and Zubakov (1962, Sections 13 and Appendix III), and Pinsker (1964, pp. 164, 170, 172).

## 8.8.2 Summary of Prediction

Strictly stationary processes with a constant variance $\sigma^2$ are considered. Then

$$\sigma^2 \geq e(\infty) \geq e(\ell) \geq e(0) = 0 \quad \text{for all } \ell \geq 0.$$

In Section 8.8 we have considered the following cases.

(i) $e(\ell) > 0$ for all $\ell > 0$.

(ia) Non-Deterministic Ergodic Processes, i.e.,

$$\sigma^2 = e(\infty) > 0 .$$

(lb) Partially Deterministic Processes, i.e.,

$$\sigma^2 > e(\infty) > 0 .$$

(ii) Completely Deterministic Processes, i.e.,

$$e(\ell) = 0 \quad \text{for all } \ell \geq 0 .$$

When ergodic this case results in perpendicular probability measures.

## Linear Prediction

Covariance stationary processes with zero means and absolutely continuous spectral distribution functions are considered. Here we have the following cases.

(i) Gaussian Ergodic Processes.

These processes always have the property that $e_L(\ell) = e(\ell)$ for all $\ell \geq 0$.

(ii) Regular Processes, i.e.,

$$e_L(\ell) > 0 \quad \text{for all } \ell > 0.$$

(iii) Singular Processes, i.e.,

$$e_L(\ell) = 0 \quad \text{for all } \ell \geq 0 .$$

## Relationships

Strictly stationary processes with zero mean and absolutely continuous spectral distribution functions are assumed for the following relationships.

(i) $e_L(\ell) \geq e(\ell)$ for all $\ell \geq 0$.

(ii) Non-deterministic or partially deterministic processes are regular.

(iii) Singular processes are completely deterministic.

(iv) Most non-gaussian completely deterministic processes are regular.

## 8.9 Gaussian Processes with Proportional Covariance Functions

### 8.9.1 Discriminating between two Gaussian Processes with Proportional Covariance Functions

Grenander (1950, p. 221) proved that perfect discrimination always arises when testing between the two hypotheses

$$H_1 : E_1[X(t)] = 0;\ E_1[X(s)X(t)] = \gamma(s,t),\ s,\ t \text{ in } T$$

and

$$H_2 : E_2[X(t)] = 0;\ E_2[X(s)X(t)] = k\gamma(s,t),\ s,\ t \text{ in } T$$

where K is a constant, and where $X(t)$ is a gaussian process, assuming that the multidimensional sample observations $\{X(t_i)\}$ ($t_i$ in T, $i = 1, \ldots, n$) always have non-singular covariance matrices. This last condition was shown to always be true for covariance stationary random processes with a spectral density function that is not identically zero. [See Good and Doog (1958, 1959).] It follows that $W_T(H_1/H_2) = \infty$, $J_T(1,2) = \infty$, and $P_{1T} \perp P_{2T}$ from Sections 7.7, 7.8, and 7.9.

## 8.9.2 Perfect Signal Detection when the Gaussian Signal and Noise have Proportional Covariance Functions

Suppose one is testing between the two hypotheses that $R(t) = S(t) + N(t)$, t in T and $R(t) = N(t)$ (t in T) as in Section 7.9.1. Suppose also that $S(t)$ and $N(t)$ are independent stationary gaussian processes with zero means and $\gamma_S(\tau) = \alpha\gamma_N(\tau)$, where $\alpha = \sigma_S^2/\sigma_N^2$. Then

$$\gamma_{S+N}(\tau) = \gamma_S(\tau) + \gamma_N(\tau) = (\alpha + 1)\gamma_N(\tau)$$

as shown by (4.6.3.1). If $S(t)$ and $N(t)$ have absolutely continuous spectral distribution functions then it follows from Section 8.9.1 that $P_{\mathbf{N}_T} \perp P_{\mathbf{S}_T + \mathbf{N}_T}$ and $W_T(H_N/H_{S+N}) = \infty$, and $J_T(N, S+N) = \infty$. It also follows from Baker (1969a) that $I(\mathbf{R}_T : \mathbf{S}_T) = \infty$ as will be shown in Section 11.1.

One way in which $\gamma_S(\tau) = \alpha\gamma_N(\tau)$ would be if $S(t)$ and $N(t)$ are both band-limited white noise over the same frequency band and fall off at the same rate. This follows directly from (8.4.1) where $\alpha = (\sigma_N^2/\sigma_N^2)$. This was shown by Good and Doog in 1958.

It will now be shown directly that $I(\mathbf{R}_T : \mathbf{S}_T) = \infty$, $W_T(H_N/H_{S+N}) = \infty$, and

$J_T(N, S + N) = \infty$ in the same manner that was used by Good in 1958 and 1960.

## 8.9.3 $I(\mathbf{R}_T : \mathbf{S}_T)$ when the Gaussian Signal and Noise have Proportional Covariance Functions

Let S(t) and N(t) be defined as in Section 8.9.2. Define $T_n = \{t_i\}$ ($t_i$ in T, i = 1, ..., n); $\mathbf{S}_n = \{S(t_i)\}$; $\mathbf{N}_n = \{N(t_i)\}$; $\mathbf{R}_n = \{R(t_i)\}$; ($t_i$ in $T_n$). Also define $\mathbf{S}_T = \{S(t)\}$; $\mathbf{N}_T = \{N(t)\}$; $\mathbf{R}_T = \{R(t)\}$; (t in T). Then from Section 4.6.3, where $\mathbf{I}_n$ is the nth order identity matrix,

$$I(\mathbf{R}_n : \mathbf{S}_n) = \frac{1}{2} \log|\mathbf{I}_n + \mathbf{A}_{\mathbf{S}_n} \mathbf{A}_{\mathbf{N}_n}^{-1}| = \frac{1}{2} \log|(1 + \alpha)\mathbf{I}_n| = \frac{n}{2} \log(1 + \alpha) \qquad (8.9.3.1)$$

where $\alpha = \sigma_S^2/\sigma_N^2$ .

Therefore, from Section 7.2 and the fact that the covariance matrices are non-singular, as pointed out in Section 8.9.1,

$$I(\mathbf{R}_T : \mathbf{S}_T) = \lim_{T_n \to T} I(\mathbf{R}_n : \mathbf{S}_n) = \lim_{n\to\infty} \frac{n}{2} \log (1 + \alpha) = \infty$$

for all T. Thus, from Section 7.8, $P_{\mathbf{R}_T \mathbf{S}_T} \perp P_{\mathbf{R}_T} \times P_{\mathbf{S}_T}$ . The expected mutual information rate, which may be defined as

$$\bar{I} = \lim_{T\to\infty} \frac{1}{T} I(\mathbf{R}_T : \mathbf{S}_T)$$

is also infinite in this case. Good and Doog (1958) were the first to notice this "paradox". Shannon (1948) overlooked it by sampling only at the uncorrelated points (see Section 8.4) at a constant distance apart equal to 1/2W. The expected mutual information at each point is then $\frac{1}{2} \log (1 + \alpha)$, and the mutual information rate is then

$$[\frac{1}{2} \log (1 + \alpha)]/(1/2W) = W \log (1 + \alpha) \qquad (8.9.3.2)$$

[Shannon (1948, p. 103)]. The expected mutual information over an interval T at this rate would be

$$T\,W \log (1 + \alpha) \tag{8.9.3.3}$$

which would always be finite for finite T. Formula (8.9.3.3) was very frequently mentioned in the engineering literature in the decade following Shannon's 1948 paper.

### 8.9.4 $W_T(H_N/H_{S+N})$ and $J_T(N, S+N)$ when the Gaussian Signal and Noise have Proportional Covariance Functions

Let S(t) and N(t) be defined as in Section 8.9.2. Then from (5.6.1),

$$W_n(H_{S+N}/H_N) = -\frac{1}{2}\log|(1+\alpha)I_n| + \frac{1}{2}\operatorname{tr}(\alpha I_n)$$

$$= -\frac{n}{2}\log(1+\alpha) + \frac{n}{2}\alpha = \frac{n}{2}[\alpha - \log(1+\alpha)].$$

Similarly, from (5.7.2),

$$W_n(H_{S+N}/H_N|H_N) = -\frac{1}{2}\log|I_n + \alpha I_n| + \frac{1}{2}\operatorname{tr}[I_n - (\alpha I_n + I_n)^{-1}]$$

$$= -\frac{n}{2}\log(1+\alpha) + \frac{1}{2}\operatorname{tr}\,(1 - \frac{1}{\alpha+1})I_n$$

$$= \frac{n}{2}[\frac{\alpha}{\alpha+1} - \log(1+\alpha)] .$$

These results are from Good (1960, p. 119). It also follows from (6.5.1) that

$$J_n(S+N,\, N) = \frac{n}{2}[\alpha - \log(1+\alpha)] - \frac{n}{2}[\frac{\alpha}{\alpha+1} - \log(1+\alpha)] = \frac{n\alpha^2}{2(\alpha+1)} .$$

Therefore from Sections 7.4 and 7.5 it follows that

$$W_T(H_{S+N}/H_N) = \lim_{n\to\infty} W_n(H_{S+N}/H_N) = \infty ,$$

$$W_T(H_{S+N}/H_N|H_N) = \lim_{n\to\infty} W_n(H_{S+N}/H_N|H_N) = \infty ,$$

and

$$J_T(S+N, N) = \lim_{n\to\infty} J_n(S+N, N) = \infty$$

for all T. It can also be concluded from Sections 7.7 and 7.8 that $P_{S_T+N_T} \perp P_{N_T}$. Also

$$\overline{W} = \lim_{T\to\infty} \frac{1}{T} W_T(H_{S+N}/H_N) = \infty$$

and

$$\overline{J} = \lim_{T\to\infty} \frac{1}{T} J_T(S+N, N) = \infty.$$

### 8.9.5 When $S_T$ and $N_T$ are both Periodic Gaussian Band-Limited White Noise

Suppose S(t) and N(t) are both periodic gaussian band-limited white noise over the same frequency band W. Then the spectral distribution function is not absolutely continuous, but discrete, as pointed out in Section 8.6.

The rank of the covariance matrices of $S(t_i)$ and $N(t_i)$ ($t_i$ in T) (i = 1, ..., n) is never greater than $2n_o - 1$, where $n_o$ is defined in Section 8.6. (See Good and Doog 1958 and 1959.) It therefore follows from (4.6.3), (5.6.1), (5.7.2) and (6.5.1), and Sections 7.2, 7.4, and 7.5 that

$$I(R_T : S_T) = \frac{2n_o-1}{2} \log (1 + \alpha) < \infty , \qquad (8.9.5.1)$$

where $\alpha = \sigma_S^2/\sigma_N^2$, and

$$W_T(H_{S+N}/H_N) = \frac{2n_o-1}{2}[\alpha - \log (1 + \alpha)] < \infty ,$$

$$W_T(H_{S+N}/H_N|H_N) = \frac{2n_o-1}{2} [\frac{\alpha}{\alpha+1} - \log (1 + \alpha)] < \infty ,$$

and

$$J_T(S, S+N) = (2n_o -1)\alpha^2/2(\alpha + 1) < \infty .$$

It still follows that the total amount of expected mutual information, expected weight of evidence, and the divergence, which are all finite, may be found by sampling at any $2n_o$-1 points in any arbitrarily small interval of T, even in this case.

Since $\frac{n_o - 1}{T} \leq W < \frac{n_o}{T}$, it follows that, for a fixed W, $n_o$ must increase as T increases, so that, for large values of T, $n_o$ is approximately equal to TW. Therefore, it follows from (8.9.5.1) that $I(R_T : S_T)$ is approximately equal to (TW - 0.5) log (1 + α), and therefore, the expected mutual information rate is

$$\bar{I} = \lim_{T\to\infty} \frac{1}{T} I_T(R_T : S_T) = \lim_{T\to\infty} (W - \frac{1}{2T}) \log (1 + \alpha) = W \log (1 + \alpha)$$

which is the same as Shannon's result in Section 8.9.3, as pointed out by Good and Doog (1958, 1959). This is because $I(R_T : S_T)$ is completely determined, for the periodic case, by sampling at the $2n_o$-1 uncorrelated points, as pointed out in Section 8.6.

# IX. OTHER EXPRESSIONS FOR EXPECTED MUTUAL INFORMATION

In this chapter the expected mutual information is expressed in terms of functionals, Hilbert space projection operators, and integral operators, and includes improper generalized processes such as white noise. Most of the chapter is restricted to gaussian processes. Sufficient conditions are given in Sections 9.2.2.2 and 9.2.5 for $I(X_Y:Y_S)$ to be finite. In Section 9.2.3.1 sufficient conditions are given for $I(R_T:S_T)$ to be finite when the gaussian noise is white. It is pointed out in Section 9.2.3.2, when $S_T$ and $N_T$ are both gaussian white noise, and in Section 9.4 when they are stationary gaussian processes with proportional spectral density functions, that $I(R_T:S_T)$ is infinite.

In Section 7.2 $I(X_T:Y_S)$ was defined as the Riemannian limit of $I(X_n:Y_m)$. It is often difficult to use such a definition and moreover it does not include white noise, as defined in Section 8.5.

## 9.1 Expression of Expected Mutual Information in Terms of Functionals

The technique of generalized function theory will now be used. For a reference see Gelfand and Yaglom (1959, p. 210).

Let $\Phi$ and $\Psi$ be sets of infinitely differentiable functions on closed and bounded intervals T and S, respectively. Assume that these functions vanish outside these intervals. For each $\phi$ in $\Phi$ and $\psi$ in $\Psi$ it is assumed that the processes are such that

$$X(\phi) = \int_0^T X(t)\phi(t)dt$$

and

$$Y(\psi) = \int_0^S Y(s)\psi(s)ds$$

are well-defined ordinary random variables. Observe that

$$E[X(\phi)] = \int_0^T E[X(t)]\phi(t)dt,$$

$$E[X^2(\phi)] = \int_0^T\int_0^T E[X(t)X(u)]\phi(t)\phi(u)dt\,du,$$

and

$$E[X(\phi)Y(\psi)] = \int_0^S \psi(s)ds \int_0^T E[X(t)Y(s)]\phi(t)dt\ .$$

If X(t) is white noise with $E[X(t)] = 0$, then

$$\sigma^2_{X(\phi)} = K\int_0^T \phi^2(t)dt$$

as shown in Section 8.5.

Gelfand and Yaglom (1959, p. 211) have defined the expected mutual information by

$$I(X_T:Y_S) = \sup I\{[X(\phi_1), \ldots, X(\phi_k)]:[Y(\psi_1), \ldots, Y(\psi_\ell)]\} \tag{9.1.1}$$

where the sup is over all $\phi_i$ in $\Phi$ and $\psi_j$ in $\Psi$ for all integers k and $\ell$. This definition agrees with the Riemannian definition whenever both these definitions are meaningful.

## 9.2 Gaussian Processes

### 9.2.1 Expression of Expected Mutual Information in Terms of Correlation Coefficients of Functionals

Let $\Phi' = \{\phi_i\}(i = 1, 2, \ldots)$ and $\Psi' = \{\psi_j\}(j = 1, 2, \ldots)$ be an orthonormal (not necessarily complete) system of infinitely differentiable functions on T and S which vanish outside these closed and bounded intervals. Then

$$\int_0^T \phi_i(t)\phi_j(t)dt = \int_0^S\int_0^T \phi_i(t)\psi_j(s)dt\,ds = \int_0^S \psi_i(s)\psi_j(s)ds = \delta_{ij}$$

where $\delta_{ij} = 0$ if $i \neq j$ and 1 if $i = j$. Assume also that

$$E[X(\phi_i)X(\phi_j)] = E[X(\phi_i)Y(\psi_j)] = E[Y(\psi_i)Y(\psi_j)] = 0$$

if $i \neq j$.

When $E[X(t)] = E[Y(s)] = 0$, it follows from (4.6.0.3) and Galfand and Yaglom (1959, p. 217) that

$$I(X_T:Y_S) = -\frac{1}{2}\sum_{j=1}^{n} \log (1 - \rho_j^2) \qquad (9.2.1)$$

where n is the smaller of the two numbers of functions in $\Phi'$ or $\Psi'$, which may be infinite. The correlation coefficient of $X(\phi_j)$ and $Y(\psi_j)$ $(j = 1, 2, \ldots)$ is $\rho_j$. It may be observed that $-\log (1 - \rho_j^2)$ is non-negative for $j = 1, 2, \ldots$ .

## 9.2.2 Expression of Expected Mutual Information in Terms of Hilbert Space Projection Operators

### 9.2.2.1 Definitions

Let $\Phi$ and $\Psi$ be defined as in Section 9.1. Let $X(\phi)$ ($\phi$ in $\Phi$) and $Y(\psi)$ ($\psi$ in $\Psi$) represent random variables with finite variances defined as in Section 9.1. Let $H_X$ represent the Hilbert space consisting of all h such that there exists constants $\{c_i\}$ and random variables $\{X(\phi_i)\}$ ($\phi_i$ in $\Phi$) satisfying

$$\lim_{m\to\infty} E[h - \sum_{i=1}^{m} c_i X(\phi_i)]^2 = 0 .$$

Let $H_Y$ be similarly defined for $Y(\psi_i)$ ($\psi_i$ in $\Psi$). Define H to be a Hilbert space of random variables with finite variances that contains $H_X$ and $H_Y$. The inner product in these Hilbert spaces for any two elements h and g in H is defined as E(hg) and therefore the norm is $||h|| = \sqrt{E(h^2)}$. Complete Hilbert spaces also have the property that if any sequence of variables $h_i$ $(i = 1, 2, \ldots)$ in the space is such that

$$\lim_{\substack{n\to\infty \\ m\to\infty}} ||h_n - h_m|| = 0$$

then

$$\lim_{n\to\infty} ||h_n - h|| = 0$$

for some h in the space. Every element h in H can be written as a sum $h_X^* + h_X$ where $h_X$ is in $H_X$ and $E(h_X^* g) = 0$ for all g in $H_X$. Then, of course, $h_X^*$ can be expressed as $h_X^* = h - h_X$. This is similar to the definitions in Sections 4.6.1 and 4.6.3.

The projection operator $\mathcal{E}_X$ of H onto $H_X$ is defined for any h in H by $\mathcal{E}_X h = h_X$. It is linear since

$$\mathcal{E}_X(ah + bg) = ah_X + bg_X = a\mathcal{E}_X h_X + b\mathcal{E}_X g$$

and idempotent since

$$\mathcal{E}_X \mathcal{E}_X h = \mathcal{E}_X h_X = h_X = \mathcal{E}_X h$$

and self-adjoint since

$$E[(\mathcal{E}_X h) g] = E[h(\mathcal{E}_X g)]$$

because

$$E[(\mathcal{E}_X h) g] = E[h_X(g_X^* + g_X)] = E[h_X g_X]$$

and

$$E[h(\mathcal{E}_X g)] = E[(h_X^* + h_X) g_X] = E[h_X g_X] .$$

$||\mathcal{E}_X||$ is defined as inf K where

$$||\mathcal{E}_X h|| = \sqrt{E(h_X^2)} < K||h||$$

where $||h|| = \sqrt{E(h^2)}$. $h_X^*$ may be expressed by

$$h_X^* = h - h_X = h - \mathcal{E}_X h = (I - \mathcal{E}_X)h .$$

Similarly, $\mathcal{E}_Y$ is defined as the projection operator of H onto $H_Y$. For a further discussion see, for instance, Riesz and Nagy (1956).

### 9.2.2.2 Expected Mutual Information

The results of this section are from Gelfand and Yaglom (1959, p. 221). Define the symmetric operators $B_1 = E_X E_Y E_X$ and $B_2 = E_Y E_X E_Y$. Then a necessary and sufficient condition for the expected mutual information to be finite is that either $B_1$ or $B_2$ is compact with finite trace. An operator $B_1$ is said to be compact in H (also called completely continuous) if every infinite sequence $\{h_n\}(n = 1, 2, \ldots)(h_n \text{ in } H)$, $||h_n|| < C$ for some constant C, has the property that the sequence $\{B_1 h_n\}$ contains a subsequence $\{B_1 h_{n_k}\}$ such that there exists an h in H where

$$\lim_{n_k \to \infty} ||B_1 h_{n_k} - h|| = 0 .$$

The operator $B_1$ is said to have finite trace if the sum of its eigenvalues is finite; that is, if $\sum_{j=1}^{n} \lambda_{1j}^2 < \infty$ where $B_L h = \lambda_{ij}^2 h$, and n may be infinity. Here and throughout the rest of this monograph these eigenelements h represent an orthonormal basis for H. Each eigenvalue $\lambda_{1j}$ occurs as many times in the sum as there are orthonormal eigenelements corresponding to that eigenvalue. When $B_1$ is compact with finite trace then so is $B_2$ and

$$I(X_T:Y_S) = -\frac{1}{2}\sum_{j=1}^{n} \log(1 - \lambda_{1j}^2) = -\frac{1}{2}\sum_{j=1}^{n} \log(1 - \lambda_{2j}^2)$$

$$= -\frac{1}{2}\log|I - B_1| = -\frac{1}{2}\log|I - B_2| \qquad (9.2.2.2.1)$$

where n may be infinity and where the determinant $|I - B_1|$ is defined as the product of its eigenvalues, that is as $\prod_{j=1}^{n}(1 - \lambda_{1j}^2)$ . It is shown by Gelfand and Yaglom (1959, p. 219) that $\rho_j^2 = \lambda_{1j}^2 = \lambda_{2j}^2$ $(j = 1, 2, \ldots)$ for the correlation coefficients $\rho_j^2$ in Equation (9.2.1), and therefore $|\lambda_{1j}^2| \leq 1$ and $|\lambda_{2j}^2| \leq 1$.

### 9.2.3 $I(R_T:S_T)$ when the Gaussian Noise is White

Suppose $R(t) = S(t) + N(t)$ (t in T), where S(t) and N(t) are independent gaussian processes with zero means, and suppose, throughout Section 9.2.3, that

N(t) is white noise with constant spectral density function K.

### 9.2.3.1 An Orthonormal System of Functions

Suppose that the covariance function $\gamma_S(s,t)$ of the signal is continuous and finite over the closed interval T. Let $\Phi = \{\phi_i\}$ be an orthonormal system (not necessarily complete) of eigenfunctions of the kernel $\gamma_S(s,t)$. That is,

$$\int_T \gamma_S(s,t)\phi_i(s)ds = \lambda_i^2 \phi_i(t) . \qquad (9.2.3.1.1)$$

Any covariance function will always have non-negative eigenvalues. [Davenport and Root (1958, p. 373).] It then follows (if $S(\phi_i)$ and $S(\phi_j)$ are defined as in Section 9.1) that

$$E[S(\phi_i)S(\phi_j)] = \int_0^T \phi_i(s)ds \int_0^T \gamma_S(s,t) \ \phi_j(t)dt$$

$$= \lambda_j^2 \int_0^T \phi_i(s)\phi_j(s)ds = \lambda_j^2 \delta_{ij} ,$$

and therefore $\{S(\phi_j)\}(\phi_j$ in $\Phi)$ is an orthogonal system in the Hilbert space $H_S$. (An orthogonal system in any Hilbert space H is any set of elements in H such that the inner product of any two distinct elements in the set is zero.) It also follows from the definition of white noise that $\gamma_N(t-s) = K\delta(t-s)$. Hence

$$E\{[S(\phi_i) + N(\phi_i)][S(\phi_j) + N(\phi_j)]\} = (\lambda_j^2 + K)\delta_{ij} ,$$

and therefore $R(\phi_j) = S(\phi_j) + N(\phi_j)(\phi_j$ in $\Phi)$ is an orthogonal system in the Hilbert space $H_R$. Similarly,

$$E\{S(\phi_i)[S(\phi_j) + N(\phi_j)]\} = E[S(\phi_i)S(\phi_j)] = \lambda_j^2 \delta_{ij} .$$

Therefore, it follows from (9.2.1) that

$$I(\mathbf{R}_T:S_T) = -\frac{1}{2}\sum_{j=1}^{N} \log(1-\rho_j^2)$$

where n may be infinity and

$$\rho_j^2 = \frac{E^2[S(\phi_j)R(\phi_j)]}{E[S^2(\phi_j)]E[R^2(\phi_j)]} = \frac{\lambda_j^4}{\lambda_j^2(\lambda_j^2+K)} = \frac{\lambda_j^2}{\lambda_j^2+K}$$

so that

$$I(\mathbf{R}_T:S_T) = -\frac{1}{2}\sum_{j=1}^{N}\log(1-\frac{\lambda_j^2}{\lambda_j^2+K}) = \frac{1}{2}\sum_{j=1}^{N}\log(1+\frac{\lambda_j^2}{K}) < \frac{1}{2K}\sum_{j=1}^{N}\lambda_j^2 ,$$

where N may be infinity.

Root (1968, p. 161) and Balakrishnan (1968, p. 205) show that $I(\mathbf{R}_T:S_T)$ will always be finite when $S(t)$ is a stationary gaussian process with a continuous covariance function and a continuous spectral density function that is not identically zero, when the gaussian noise is white. It will be shown in Section 11.1 that $W_T(H_{S+N}/H_N)$ will also be finite in this case.

When the sum of the eigenvalues in (9.2.3.1.1) are finite, the projection operator also has finite trace (defined in Section 9.2.2.2) because $I(\mathbf{R}_T:S_T)$ is finite.

The expected mutual information can be expressed in terms of the covariance function of the signal by using the following expansion pointed out by Balakrishnan (1968, p. 205). If $|\lambda_j^2| < K$ $(j = 1, 2, \ldots)$ then the above expression for $I(\mathbf{R}_T:S_T)$ becomes

$$I(\mathbf{R}_T:S_T) = \frac{1}{2}\sum_{j=1}^{N}\log\,(1+\frac{\lambda_j^2}{K}) = \frac{1}{2}\sum_{n=0}^{\infty}(-1)^n\sum_{j=1}^{N}\frac{(\lambda_j^2/K)^{n+1}}{n+1}$$

$$= \frac{1}{2}\sum_{n=0}^{\infty}\frac{(-1)^n}{(n+1)K^{n+1}}\sum_{j=1}^{N}(\lambda_j^2)^{n+1}$$

where

$$\sum_{j=1}^{N} (\lambda_j^2)^{n+1} = \int_0^T \cdots \int_0^T \gamma_S(t_1,t_2)\gamma_S(t_2,t_3)\cdots \gamma_S(t_{n+1},t_1)dt_1\cdots dt_{n+1} \ .$$

### 9.2.3.2 $I(R_T:S_T)$ when $S_T$ and $N_T$ are both Gaussian White Noise

If S(t) is also white noise with spectral density $C > 0$, then from the previous section,

$$\int_T \gamma_S(t,s)\phi_i(s)ds = C\int_T \delta(t-s)\phi_i(s)ds = C\ \phi_i(t)$$

and

$$I(R_T:S_T) = \frac{1}{2}\sum_{j=1}^{\infty} \log(1 - \frac{C}{K}) = \infty \ .$$

The reason there are a denumerable number of terms in the sum is because the orthonormal set $\Phi = \{\phi_i\}$ is denumerable when the eigenvalues $\lambda_i^2 = C$ are all positive. See, for instance, Courant and Hilbert (1953, Vol. I., Chap. 3, Art. 4 and 5) and Riesz and Nagy (1956, Chap. I., p. 67) for the above reasoning and the following definitions. The result above is similar to the result in Section 8.9.3 since the signal and noise have proportional spectral density functions C and K. These results can now be generalized by using the following definitions.

### 9.2.3.3 Definitions

(i) $L_2(T) = \{f(t)\,|\,(t \text{ in } T) \text{ and } \int_T |f(t)|^2 dt < \infty\}$

$L_2 \quad = \{f(t)\,|\int_{-\infty}^{\infty} |f(t)|^2 dt < \infty\}$ .

(ii) $\gamma(t,s)$ is positive definite if

$$\int_T\int_T \gamma(t,s)g(s)g(t)ds\ dt > 0$$

for every function $g(t)$ in $L_2(T)$ such that

$$\int_T |g(t)|^2 dt > 0 \ .$$

For instance, in Section 9.2.3.1, if $\gamma_S(s,t)$ is positive definite, then

$$E[S^2(\phi_i)] = \iint \gamma_S(t-s)\phi_i(s)\phi_i(t)ds\ dt = \lambda_i^2 > 0$$

provided $\int_T |\phi_i(t)|^2 dt > 0$ .

(iii) Complete System

$\Phi$ is complete in $L_2(T)$ if for any $f(t)$ in $L_2(T)$ there exists a set of constants

$$\alpha_i = f(\phi_i) = \int_T f(t)\phi_i(t)dt \quad (\phi_i \text{ in } \Phi)\ (i = 1, 2, \ldots)$$

such that

$$\lim_{n\to\infty} \int_T [f(t) - \sum_{i=1}^{n} \alpha_i\phi_i(t)]^2 dt = 0 .$$

This last expression is abbreviated by saying that

$$f(t) = \ell.\text{i.m.} \sum_{i=1}^{n} \alpha_i\phi_i(t)$$

where $\ell$.i.m. stands for the limit in mean square.

## 9.2.4 $I(R_T:S_T)$ when the Gaussian Signal and Noise have Proportional Covariance Functions

The results of Section 8.9.3 may now be seen by using these definitions. Suppose that $S(t)$ and $N(t)$ are independent gaussian processes with zero means and covariance functions satisfying $\gamma_S(t,s) = \alpha\ \gamma_N(t,s)$ where $\alpha = \sigma_S^2(t)/\sigma_N^2(t)$. Let $\Phi$ be defined as in Section 9.2.3.1. Then

$$E[S^2(\phi_j)] = E[S(\phi_j)R(\phi_j)] = \lambda_j^2$$

and

$$\begin{aligned} E[R^2(\phi_j)] &= E[S^2(\phi_j)] + E[N^2(\phi_j)] \\ &= \lambda_j^2 + \int_0^T\int_0^T \gamma_N(t,s)\phi_j(s)\phi_j(t)dt\ ds \\ &= \lambda_j^2 + (\lambda_j^2/\alpha) \end{aligned}$$

because $\gamma_N(t,s) = \gamma_S(t,s)/\alpha$. Therefore,

$$\rho_j^2 = \lambda_j^4/[\lambda_j^2 \; \lambda_j^2 \; (1 + \frac{1}{\alpha})] = \alpha/(\alpha + 1)$$

and

$$I(R_T:S_T) = -\frac{1}{2} \sum_{j=1}^{N} \log(1 - \frac{\alpha}{\alpha+1}) = \frac{N}{2} \log(\alpha + 1) .$$

$\Phi$ is a complete denumerable orthonormal system in $L_2(T)$ provided $\gamma_S(t,s) = \alpha \; \gamma_N(t,s)$ is positive definite, and in that case it follows from Equation 9.2.1 that $I(R_T:S_T) = \infty$ because there are a denumerable number of eigenfunctions in $\Phi$. $\gamma_S(t,s)$ will always be positive definite when $S(t)$, and therefore $N(t)$, are stationary gaussian processes with a spectral density function that is not identically zero, as pointed out by Root (1968, p. 190). These results therefore agree with Good and Doog (1958, 1959) as mentioned in Sections 8.9.1 and 8.9.3. These results are also mentioned in Balakrishnan (1968, p. 206).

### 9.2.5 Expression of Expected Mutual Information in Terms of Integral Operators

These results are from Yaglom (1962, p. 333), Gelfand and Yaglom (1959, p. 232), and Baker (1969a, p. 1). Let $X_T$ and $Y_S$ be defined as in Section 7.1. Let f be any function in $L_2(T)$. Define $B_{XX}$ as the integral operator with the covariance kernel $\gamma_X(s,t)$ so that

$$B_{XX} \; f(t) = \int_0^T \gamma_X(s,t) f(t) dt$$

where

$$B_{XX}^{-1} \; B_{XX} \; f(t) = f(t) .$$

Similarly, let $B_{YY}$, $B_{XY}$, and $B_{YX}$ be defined as the integral operators with kernels $\gamma_Y$, $\gamma_{XY}$, and $\gamma_{YX}$, respectively. For instance, if w(s) is any function in $L_2(S)$ then

$$B_{XY}\ w(s) = \int_0^S \gamma_{XY}(t,s)w(s)ds\ .$$

Now define

$$B = B_{XY}\ B_{YY}^{-1}\ B_{YX}\ B_{XX}^{-1}\ .$$

Then $I(X_T:Y_S)$ is finite if B is compact with a finite trace, just like the situation in Section 9.2.2.2 for projection operators. In this case

$$I(X_T:Y_S) = -\frac{1}{2}\sum_{j=1}^{N} \log\ (1 - \lambda_j^2)$$

where $\lambda_j^2 = \rho_j^2$ $(j = 1, 2, \ldots)$ are the eigenvalues of B.

## 9.2.6 Radon-Nikodym Derivative

As pointed out in Section 4.5.1 if $I(X_T:Y_S)$ is finite then $P_{X_TY_S}$ is absolutely continuous with respect to $P_{X_T} \times P_{Y_S}$ and from Section 7.2,

$$I(X_T:Y_S) = \iint_{\Omega_T\times\Omega_S} \log \frac{dP_{X_TY_S}}{dP_{X_T}\times P_{Y_S}}\ dP_{X_TY_S}$$

where $\dfrac{dP_{X_TY_S}}{dP_{X_T}\times P_{Y_S}}$ is the Radon-Nikodym derivative, as defined in Section 3.1.1.

When $I(X_T:Y_S)$ is finite, the Radon-Nikodym derivative can be found in terms of the eigenvalues $\lambda_i$ of B. In this case the Radon-Nikodym derivative is

$$\prod_{i=1}^{N} (1-\lambda_i^2)^{-\frac{1}{2}} \exp\left\{\frac{1}{2}\sum_{i=1}^{N} \frac{\lambda_i^2[X^2(\phi_i)+Y^2(\psi_i)-2\lambda_i X(\phi_i)Y(\psi_i)]}{1-\lambda_i^2}\right\}$$

where $\Phi = \{\phi_i\}$ and $\Psi = \{\psi_i\}$ are defined as in Section 9.2.1.

## X. EXPRESSING EXPECTED WEIGHT OF EVIDENCE FOR GAUSSIAN GENERALIZED PROCESSES IN TERMS OF INTEGRAL OPERATORS

In this chapter the expected weight of evidence for gaussian continuous-parameter or generalized processes, such as white noise, is expressed in terms of integral operators. Necessary and sufficient conditions are given for $P_{1T}$ and $P_{2T}$ to be parallel or perpendicular in terms of integral operators with covariance kernels. The relationship between strong parallelism and the Radon-Nikodym derivative is given and applied to joint and product gaussian probability measures and stationary gaussian processes with rational spectral density functions. The results of this chapter are from Yaglom (1962, pp. 334-337).

Let $\mathbf{X}_{1T} = \{X_1(t)\}$ (t in T) and $\mathbf{X}_{2T} = \{X_2(t)\}$ (t in T) represent continuous or generalized (to include white noise) gaussian processes. Then from Section 7.8 $P_{1T}$ and $P_{2T}$ are either parallel or perpendicular depending on whether $W_T(H_1/H_2)$ is finite or infinite. $W_T(H_1/H_2)$ is defined similarly to (9.1.1). $P_{1T}$ is parallel to $P_{2T}$ if and only if

$$B = I - B_{11}^{-1} B_{22}$$

is a compact Hilbert-Schmidt operator, where $B_{11}$ and $B_{22}$ are integral operators with covariance kernels, defined similarly to those in Section 9.2.5. That is, if f is in $L_2(T)$, then

$$B_{11}f(t) = \int_0^T \gamma_1(s,t)\ f(t)\ dt\ .$$

A compact operator has the same meaning as in Section 9.2.2.2. An operator is called a Hilbert-Schmidt operator if it has the property that the sum of the squares of its eigenvalues is finite. Thus, B is a Hilbert-Schmidt operator if

$$\sum_{i=1}^{\infty} (1 - \lambda_i^2)^2 < \infty$$

where $\lambda_i^2$ are the eigenvalues of $B_{11}^{-1} B_{22}$ .

## 10.1 Strongly Parallel Probability Measures

If B is compact and has finite trace (i.e. $\sum_{i=1}^{\infty} |1 - \lambda_i^2| < \infty$), then $P_{1T}$ and $P_{2T}$ are said to be strongly parallel.

Theorem 10.1. Strong parallelism implies parallelism.

Proof: Assume strong parallelism. Then B has finite trace. Decompose the trace as follows:

$$\sum_{i=1}^{\infty} |1 - \lambda_i^2| = \sum_{|1-\lambda_i^2| \geq 1} |1 - \lambda_i^2| + \sum_{|1-\lambda_i^2| < 1} |1 - \lambda_i^2| .$$

If $\sum_{i=1}^{\infty} |1 - \lambda_i^2| < \infty$ then there can only be a finite number of terms in the first sum. Observe that

$$\sum_{|1-\lambda_i^2|<1} (1 - \lambda_i^2)^2 < \sum_{|1-\lambda_i^2|<1} |1 - \lambda_i^2| .$$

Therefore, if B has finite trace then $\sum_{i=1}^{\infty} (1 - \lambda_i^2)^2 < \infty$ and B is a Hilbert-Schmidt operator. Now since B is compact it follows from the introduction to this chapter that $P_{1T} \| P_{2T}$ . Q.E.D.

Only for strongly parallel gaussian measures can the Radon-Nikodym derivatives be evaluated by considering the ratio of two n-dimensional gaussian densities and taking the limits independently as $n \to \infty$ of the factor before the exponent and of the exponential factor. There are examples where the probability measures of two gaussian processes are parallel but not strongly parallel.

## 10.2 Expressions for $W_T(H_1/H_2)$ and $\frac{dP_{2T}}{dP_{1T}}$

If $P_{1T} \parallel P_{2T}$ then

$$W_T(H_2/H_1) = \frac{1}{2}\sum_{i=1}^{\infty}(\lambda_i^2 - 1 - \log \lambda_i^2)$$

where $\lambda_i^2$ are the eigenvalues of $B_{11}^{-1} B_{22}$ and

$$\frac{dP_{2T}}{dP_{1T}} = \exp\{-\frac{1}{2}\sum_{i=1}^{\infty}[\log \lambda_i^2 + \frac{(1-\lambda_i^2)}{\lambda_i^2} x^2(\phi_i)]\},$$

where the $\phi_i(t)$ are the orthonormal eigenfunctions of $B_{11}^{-1} B_{22}$.

If $P_{1T}$ and $P_{2T}$ are strongly parallel then $W_T$ can be expressed as

$$W_T(H_1/H_2) = -\frac{1}{2}\sum_{i=1}^{\infty}(1-\lambda_i^2) + \log(\prod_{i=1}^{\infty}\lambda_i^2)^{-\frac{1}{2}}$$

and

$$\frac{dP_{2T}}{dP_{1T}} = (\prod_{i=1}^{\infty}\lambda_i^2)^{-\frac{1}{2}} \exp[-\frac{1}{2}\sum_{i=1}^{\infty}\frac{(1-\lambda_i^2)}{\lambda_i^2} x^2(\phi_i)] .$$

## 10.3 Relationship between Joint and Product Gaussian Probability Measures

If $X(t)$ ($t$ in $T$) and $Y(s)$ ($s$ in $S$) are gaussian processes and the joint probability is also gaussian then $P_{\mathbf{X}_T\mathbf{Y}_S}$ and $P_{\mathbf{X}_T} \times P_{\mathbf{Y}_S}$ must be either strongly parallel or perpendicular. [Yaglom (1962, p. 334).]

## 10.4 Relationship between two Stationary Gaussian Processes with Rational Spectral Density Functions

The probability measures corresponding to two covariance-stationary processes $X_1(t)$ and $X_2(t)$ ($t$ in $T$) with rational spectral density function $P_1^*(\omega)$ and $P_2^*(\omega)$ must be either strongly parallel or perpendicular depending on whether the condition

$$\lim_{|\omega|\to\infty} \frac{P_1^*(\omega)}{P_2^*(\omega)} = 1$$

holds or not. Yaglom (1962, p. 337) mentions that this result is from Slepian (1958) and Feldman (1960).

# XI. COMPARISON BETWEEN $I(R_T:S_T)$, $W_T(H_N/H_{S+N})$, AND $J_T(N, S+N)$ FOR GAUSSIAN SIGNALS AND NOISE

In this chapter a comparison is made between $I(R_T:S_T)$, $W_T(H_N/H_{S+N})$, and $J_T(N, S+N)$ for gaussian signals and noise. It is seen that $I(R_T:S_T)$ may be infinite even though $W_T(H_N/H_{S+N})$ and $J_T(N, S+N)$ are finite. It is also shown that this case does not arise if the gaussian signal and noise have rational spectral density functions. A necessary and sufficient condition is then given for them all to be infinite or all finite. A sufficient condition is also given for $P_{N_T}$ and $P_{S_T+N_T}$ to be perpendicular when the gaussian signal and noise do not have rational spectral density functions.

## 11.1 General Remarks

Baker (1969a) has shown that if $R(t) = S(t) + N(t)$ ($t$ in $T$) where $S(t)$ and $N(t)$ are independent gaussian continuous-parameter or generalized processes, then $I(R_T:S_T) < \infty$ if and only if $P_{N_T}$ and $P_{S_T+N_T}$ are strongly parallel. If $I(R_T:S_T) < \infty$ then it follows from Section 10.3 that $P_{R_TS_T}$ and $P_{R_T} \times P_{S_T}$ are also strongly parallel. If $I(R_T:S_T)$ is finite then $W_T(H_N/H_{S+N})$ and $J_T(N, S+N)$ are finite. If $W_T(H_N/H_{S+N})$ is infinite then $J_T(N, S+N)$ and $I(R_T:S_T)$ are infinite and $P_{N_T} \perp P_{S_T+N_T}$ and $P_{R_TS_T} \perp P_{R_T} \times P_{S_T}$. If, however, $P_{N_T}$ and $P_{S_T+N_T}$ are parallel, but not strongly parallel then $I(R_T:S_T) = \infty$ and $P_{R_TS_T} \perp P_{R_T} \times P_{S_T}$ and yet $W_T(H_N/H_{S+N})$ and $J_T(N, S+N)$ are finite.

## 11.2 Signal Detection

When $I(R_T:S_T) < \infty$ and $R(t)$, $S(t)$, and $N(t)$ are defined as in Section 11.1 then the perfect signal detection situation will not arise since $P_{N_T}$ and $P_{S_T+N_T}$

are parallel. If, however, $P_{N_T} \perp P_{S_T+N_T}$, so that perfect signal detection does arise, then $W_T(H_N/H_{S+N})$ and hence $I(R_T:S_T)$ are infinite. An example of this latter situation was pointed out in Section 8.9.2 when the signal and noise have proportional covariance functions.

## 11.3 Gaussian Signals and Noise with Rational Spectral Density Functions

If the gaussian signal and noise have rational spectral density functions then it follows from Section 10.4 that $P_{N_T}$ and $P_{S_T+N_T}$ can be only either strongly parallel or perpendicular depending on whether the condition

$$\lim_{|\omega| \to \infty} \frac{P_S^*(\omega)}{P_N^*(\omega)} = 0$$

holds, where $P_S^*(\omega)$ and $P_N^*(\omega)$ are the spectral density functions. It therefore follows from Section 11.1 that $I(R_T:S_T)$, $W_T(H_N/H_{S+N})$, $J_T(N, S+N)$ will all be finite or infinite depending on whether this condition holds.

If

$$\lim_{|\omega| \to \infty} \frac{P_S^*(\omega)}{P_N^*(\omega)} \neq 0$$

but $P_S^*(\omega)$ and $P_N^*(\omega)$ are not rational functions of $\omega$, but are bounded and asymptotically inversely proportional to an even power of $\omega$, then $P_{N_T}$ and $P_{S_T+N_T}$ are always perpendicular. This result was proved by Gladyshev (1961). It is also discussed by Yaglom (1962, p. 340) and by Vainshtein (1962, Appendix III).

# XII. EXPECTED MUTUAL INFORMATION RATE

In this chapter the expected mutual information rate is discussed with particular emphasis on stationary gaussian continuous-parameter or generalized signals and noise.

The upper bound of this rate is found for special cases, including the case where the signal and noise pass through a linear system. The channel capacity is also discussed, including the case where the signals have optical frequencies. The expected mutual information rate is then investigated for multidimensional random processes.

It is assumed that the means of all the random processes in this chapter are zero.

## 12.1 Introduction

The ideas in this introduction may be found in Pinsker (1964, pp. 68, 76, 77). If $X = \{X(t)\}$ $(-\infty < t < \infty)$ and $Y = \{Y(t)\}$ $(-\infty < t < \infty)$ are continuous-parameter or generalized strictly stationary random processes, then the expected mutual information rate of X and Y is defined as

$$\bar{I}(X : Y) = \lim_{T\to\infty} \frac{1}{T} I(X_T : Y_T)$$

where $I(X_T : Y_T)$ is defined by (9.1.1) to include the case where X or Y, or both, may be white noise.

If $X = \{X(t_i)\}$ $(i = 1, 2, \ldots)$ and $Y = \{Y(t_i)\}$ $(i = 1, 2, \ldots)$ are discrete-parameter strictly stationary random processes, then the expected mutual information rate of X and Y is defined as

$$\bar{I}(X : Y) = \lim_{n\to\infty} \frac{1}{n} I(X_n : Y_n)$$

where

$$X_n = \{X(t_i)\} \quad (i = 1, 2, \ldots n) \text{ and}$$

$$Y_n = \{Y(t_i)\} \quad (i = 1, 2, \ldots n) .$$

## 12.2 Continuous-Parameter and Generalized Stationary Gaussian Processes

The results of this section are from Pinsker (1964, pp. 181, 182).

Let X and Y be covariance-stationary generalized gaussian processes with spectral density functions $P^*_X(\omega)$ and $P^*_Y(\omega)$. Let

$$P^*_{XY}(\omega) = \int_{-\infty}^{\infty} e^{-i2\pi\tau\omega} \gamma_{XY}(\tau)d\tau .$$

$P^*_{XY}(\omega)$ may be complex. Under the conditions described below

$$\overline{I}(X : Y) = \frac{1}{2}\int_{-\infty}^{\infty} \log \frac{P^*_X(\omega)\; P^*_Y(\omega)}{P^*_X(\omega)\; P^*_Y(\omega) - |P^*_{XY}(\omega)|^2}\, d\omega \tag{12.2.1}$$

which may also be expressed as

$$\overline{I} = \int_0^{\infty} (\ldots)\, d\omega$$

because of the symmetry of the covariance function about the origin. Equation (12.2.1) is only valid for the following two circumstances

(i) If $P^*_X(\omega)$ or $P^*_Y(\omega)$ is a rational function of $\omega$. (See Section 8.2.1 for a discussion of such functions.)

(ii) If $P^*_X(\omega)$ or $P^*_Y(\omega)$ is equal to $R^*(\omega)\cdot|\psi^*(\omega)|^2$ where $R^*(\omega)$ is a rational function of $\omega$. This is equivalent to saying that X(t) or Y(t) is equal to

$$\int_{-\infty}^{\infty} \psi(t)\; R(t - u)du$$

as In Equation (8.2.9). It is also assumed that $|\psi^*(\omega)| \leq 1$ and $\int_{-\infty}^{\infty} |\log |\psi^*(\omega)||^2 d\omega < \infty$ so that $\log |\psi^*(\omega)|$ is in $L_2$.

If $P^*_X(\omega)$, $P^*_Y(\omega)$, and $P^*_{XY}(\omega)$ are rational functions of $\omega$ then it follows that $\overline{I}(X : Y)$ is finite if and only if $I(X_T : Y_T)$ is finite for all T. It may be observed that if $\overline{I}(X : Y)$ is finite then $I(X_T : Y_T)$ is finite for all $T > 0$ even if $P^*_X(\omega)$, $P^*_Y(\omega)$, and $P^*_{XY}(\omega)$ are not rational functions of $\omega$.

## 12.3 Discrete-Parameter Stationary Gaussian Processes

If X and Y are covariance-stationary gaussian processes then

$$\overline{I}(X : Y) = \frac{1}{2} \int_{-\frac{1}{2}}^{\frac{1}{2}} \log \frac{P^*_X(\omega)\ P^*_Y(\omega)}{P^*_X(\omega)\ P^*_Y(\omega) - |P^*_{XY}(\omega)|^2}\, d\omega \,, \tag{12.3.1}$$

which may also be expressed as

$$\overline{I} = \int_0^{\frac{1}{2}} (\ldots)\, d\omega \,.$$

The expression (12.3.1) is only valid if either X or Y is a regular process (as defined in Section 8.8.1).

If X and Y represent samples from a continuous-parameter stationary gaussian process, then

$$\lim_{n\to\infty} \frac{1}{n} I(X_n : Y_n)$$

would not be equal to the right-hand side of (12.3.1) even if $P^*_X(\omega)$ and $P^*_Y(\omega)$ were band-limited at 0.5, because X and Y would then both be singular processes (as defined in Section 8.8.1). See also Bartlett (1966, p. 179).

## 12.4 $\overline{I}(R : S)$ when the Signal and Noise are Independent Stationary Gaussian Processes with Zero Means

### 12.4.1 $\overline{I}(R : S)$ when the Gaussian Signal has a Rational Spectral Density Function

Let $R(t) = S(t) + N(t)$ $(-\infty < t < \infty)$ where $S(t)$ and $N(t)$ are independent stationary gaussian processes with zero means. Then the cross-covariance function is

$$\gamma_{RS}(\tau) = E\{R(t)\ S(t+\tau)\} = \gamma_S(\tau) \quad (0 < \tau < \infty)$$

and therefore $P^*_{RS}(\omega) = P^*_S(\omega)$, which is always a real-valued non-negative function of $\omega$. If $P^*_S(\omega) > 0$ and $P^*_N(\omega) > 0$ and $P^*_S(\omega)$ is a rational function of $\omega$ then it follows from Section 12.2 that

$$\overline{I}(R : S) = \int_0^\infty \log \frac{P^*_S(\omega)\ [P^*_S(\omega) + P^*_N(\omega)]}{P^*_S(\omega)\ [P^*_S(\omega) + P^*_N(\omega)] - [P^*_S(\omega)]^2}\, d\omega$$

$$= \int_0^\infty \log \left\{\frac{[P^*_S(\omega)]^2 + P^*_S(\omega)\ P^*_N(\omega)}{P^*_S(\omega)\ P^*_N(\omega)}\right\} d\omega = \int_0^\infty \log\left[\frac{P^*_S(\omega)}{P^*_N(\omega)} + 1\right] d\omega\ . \qquad (12.4.1.1)$$

This final expression is in Pinsker (1964, pp. 181-182). If $P^*_S(\omega)$ and $P^*_N(\omega)$ are both rational functions of $\omega$, then it follows from Section 12.2 that $\overline{I}(R : S)$ is finite if and only if $I(R_T : S_T)$ is finite for all T. It follows from Section 11.3 that this is true if and only if $W_T(H_N | H_{S+N})$ is finite for all T. Even if $P^*_S(\omega)$ and $P^*_N(\omega)$ are not rational functions of $\omega$ it is still true that if $\overline{I}(R : S)$ is finite then $I(R_T : S_T)$ and, therefore, $W_T(H_N | H_{S+N})$ are finite for all T as pointed out in Sections 12.2 and 11.1.

### 12.4.2 $\overline{I}(R : S)$ when the Gaussian Signal does not have a Rational Spectral Density Function

Fano (1961, p. 172) has proven that

$$\overline{I}(R : S) = \int_0^\infty \log\left(1 + \frac{P^*_S(\omega)}{P^*_N(\omega)}\right) d\omega$$

when $P^*_N(\omega) \neq 0$ even when $P^*_S(\omega)$ is not a rational spectral density function. If $P^*_N(\omega)$ is zero and $P^*_S(\omega)$ is not zero for some value of $\omega$ then $\overline{I}\ (R : S) = \infty$, which would be the case, for instance, if N were band-limited, but S were not. (This also follows from Sections 8.7.1.2 and 11.2.)

If S is band-limited to W and N is not band-limited then

$$\bar{I}(\mathbf{R} : \mathbf{S}) = \int_0^W \log \left(1 + \frac{P_S^*(\omega)}{P_N^*(\omega)}\right) d\omega \ .$$

If N and S are both band-limited to W then

$$\bar{I}(\mathbf{R} : \mathbf{S}) \neq \int_0^W \log \left(1 + \frac{P_S^*(\omega)}{P_N^*(\omega)}\right) d\omega \ ,$$

as proposed by Shannon (1949b, p. 19), but it can be undefined or infinite. For instance, if S and N are both band-limited white noise then $\bar{I}(\mathbf{R} : \mathbf{S}) = \infty$, as follows from Sections 8.9.2 and 8.9.3.

### 12.4.3 $\bar{I}(\mathbf{R} : \mathbf{S})$ when the Gaussian Noise is White

Balakrishnan (1968) uses the results of Section 9.2.3.1 to prove that

$$\bar{I}(\mathbf{R} : \mathbf{S}) = \int_0^\infty \log \left(1 + \frac{P_S^*(\omega)}{K}\right) d\omega$$

for the special case where N is white noise so that $P_N^*(\omega) = K$ for all $\omega$, and where S(t) has a continuous covariance function with finite variance and power, and where the spectral distribution function is absolutely continuous. He also needs to assume that K is large enough so that $\left|\frac{\lambda_j}{K}\right| < 1$ for $j = 1, 2, \ldots$, where the $\lambda_j$'s are the eigenvalues defined in Section 9.2.3.1.

He proves this by first showing that

$$\lim_{T\to\infty} \frac{1}{T} \int_0^T \cdots \int_0^T \gamma(t_1-t_2) \ \ldots \ \gamma(t_n-t_1)\, dt_1 \ \ldots \ dt_n = \int_{-\infty}^{\infty} [P_S^*(\omega)]^n d\omega$$

so that, from Section 9.2.3.1,

$$\bar{I}(\mathbf{R} : \mathbf{S}) = \frac{1}{2} \sum_{n=0}^{\infty} \frac{(-1)^n}{(n+1)K^{n+1}} \int_{-\infty}^{\infty} [P_S^*(\omega)]^{n+1} d\omega = \frac{1}{2} \int_{-\infty}^{\infty} \{ \sum_{n=0}^{\infty} (-1)^n [\frac{P_S^*(\omega)}{K}]^{n+1}/(n+1) \} d\omega$$

$$= \frac{1}{2} \int_{-\infty}^{\infty} \log \left(1 + \frac{P_S^*(\omega)}{K}\right) d\omega = \int_0^\infty \log \left(1 + \frac{P_S^*(\omega)}{K}\right) d\omega \ .$$

## 12.4.4 Upper Bound of $\overline{I}(R : S)$ when the Gaussian Noise is White

It follows from Sections 12.4.1 and 12.4.2 that, if $P_N^*(\omega) \neq 0$,

$$\overline{I}(R : S) = \int_0^\infty \log\left(1 + \frac{P_S^*(\omega)}{P_N^*(\omega)}\right) d\omega < \int_0^\infty \left(\frac{P_S^*(\omega)}{P_N^*(\omega)}\right) d\omega \qquad (12.4.4.1)$$

since $\log(1+x) < x$. Therefore, if $P_N^*(\omega) = K$ at least for large frequencies $\omega > \omega_0$ for some fixed $\omega_0$ then

$$\overline{I}(R : S) < \int_0^{\omega_0} \frac{P_S^*(\omega)}{P_N^*(\omega)} d\omega + \frac{1}{K}\int_{\omega_0}^\infty P_S^*(\omega)\, d\omega < \infty$$

if the signal variance

$$\sigma_S^2 = \gamma_S(0) = 2\int_0^\infty P_S^*(\omega)\, d\omega$$

is finite.

## 12.4.5 Upper Bound of $\overline{I}(R : S)$ for Special Cases

If it is not assumed that N is white noise, but

$$\frac{P_S^*(\omega)}{P_N^*(\omega)} \sim a\omega^{-(1+\varepsilon)} \qquad (a > 0)$$

for large $\omega > \omega_0$ for some fixed $\omega_0$ and $\varepsilon > 0$, then

$$\overline{I}(R : S) < \int_0^\infty \frac{P_S^*(\omega)}{P_N^*(\omega)} d\omega \sim \int_0^{\omega_0} \frac{P_S^*(\omega)}{P_N^*(\omega)} d\omega + a\int_{\omega_0}^\infty \omega^{-(1+\varepsilon)} d\omega$$

$$= \int_0^{\omega_0} \frac{P_S^*(\omega)}{P_N^*(\omega)} d\omega + \frac{a}{\omega^\varepsilon}\Big|_{\omega_0}^\infty$$

$$= \int_0^{\omega_0} \frac{P_S^*(\omega)}{P_N^*(\omega)} d\omega - \frac{a}{\omega_0^\varepsilon} < \infty .$$

Middleton (1961, p. 112) and Parzen (1962) [see Section (15.3.6)] arrived at a similar conclusion without using information theory.

## 12.5 $\overline{I}(R : S)$ when the Gaussian Signal and Noise pass through a Linear System

(i) Let

$$R(t) = (S(t) + N_1(t))F + N_2(t)$$

$$= S(t)F + N_1(t)F + N_2(t)$$

$$= S'(t) + N(t) \quad (-\infty < t < \infty) \qquad (12.5.1)$$

where F denotes an operator that represents the effect of a linear network or filter (see Section 8.2.1), $N_1$ denotes the external gaussian noise, and $N_2$ denotes the internal noise of the filter (as suggested by Michael M. Crum in private correspondence in 1962). It is assumed also that $S(t)$, $N_1(t)$, and $N_2(t)$ are independent stationary gaussian processes with zero means. If the linear filter F is assumed to be time invariant then it follows from Section 8.2.1 that

$$(S(t) + N_1(t))F = \int_{-\infty}^{\infty} [S(u) + N_1(u)]f(t - u)du \qquad (12.5.2)$$

where $f(t)$ is some impulse response function as defined in Section 8.2. It then follows from Equation (8.2.10) that the spectral density functions of $S'(t)$ and $N(t)$ are

$$P^*_{S'}(\omega) = |f^*(\omega)|^2 P^*_S(\omega)$$

and

$$P^*_N(\omega) = |f^*(\omega)|^2 P^*_{N_1}(\omega) + P^*_{N_2}(\omega) .$$

Therefore it follows from Section 12.4.2 that

$$I(R:S') = \int_0^\infty \log\left(1 + \frac{P^*_{S'}(\omega)}{P^*_N(\omega)}\right)d\omega = \int_0^\infty \log\left(1 + \frac{|f^*(\omega)|^2 P^*_S(\omega)}{|f^*(\omega)|^2 P^*_{N_1}(\omega) + P^*_{N_2}(\omega)}\right)d\omega$$

$$\leq \int^\infty \frac{P^*_S(\omega)}{P^*_{N_1}(\omega) + (P^*_{N_2}(\omega)/|f^*(\omega)|^2)}\, d\omega \ . \qquad (12.5.3)$$

For very high frequencies, a linear filter behaves like a capacity so that

$$|f^*(\omega)|^2 \sim a\omega^{-2}$$

[See, for instance, Newstead (1959, p. 111). This point came up in private communication between M. M. Crum and the junior author in May 1962.] Here, "a" is a constant depending on the filter. If

$$P^*_{N_2}(\omega) \geq |f^*(\omega)|^2 \ ,$$

for large values of $\omega$, then a sufficiently large value of $\omega$ can be found (say $\omega_0$) such that

$$\overline{I}(R:S') < \int_0^{\omega_0} \frac{P^*_S(\omega)}{P^*_{N_1}(\omega) + \dfrac{P^*_{N_2}(\omega)}{|f^*(\omega)|^2}}\, d\omega + \int_{\omega_0}^\infty P^*_S(\omega)\, d\omega \ . \qquad (12.5.4)$$

This follows from (12.5.3) since any spectral density function such as $P^*_{N_1}(\omega)$ is non-negative. If it is assumed that the signal has finite power (or variance) then $\overline{I}(R:S')$ is finite. The assumption that $P^*_{N_2}(\omega) \geq |f^*(\omega)|^2$ for large $\omega$ will always be satisfied if $P^*_{N_2}(\omega) \sim \alpha\omega^{-(2-\varepsilon)}$ for large $\omega$ when $\alpha$ and $\varepsilon$ are positive constants. In particular this will hold when $N_2$ is white noise because $P^*_{N_2}(\omega) = \alpha$ (a constant) when $\varepsilon = 2$.

(ii) If in the equation (12.5.1), $S''(t)$ is defined by $S''(t) = S(t)F + N_1(t)F$

then $R(t) = S''(t) + N_2(t)$ so that from Section 12.4.2

$$\overline{I}(R : S'') = \int_0^\infty \log(1 + \frac{P^*_{S''}(\omega)}{P^*_{N_2}(\omega)}) d\omega \leq \int_0^\infty \frac{P^*_{S''}(\omega)}{P^*_{N_2}(\omega)} d\omega . \tag{12.5.5}$$

If $P^*_{N_2}(\omega) = K_2$ when $\omega > \omega_0$, then

$$\overline{I}(R : S'') \leq \int_0^{\omega_0} \frac{P^*_{S''}(\omega)}{P^*_{N_2}(\omega)} d\omega + \frac{1}{K_2} \int_{\omega_0}^\infty P^*_{S''}(\omega)\, d\omega . \tag{12.5.6}$$

Observe that

$$2 \int_0^\infty P^*_{S''}(\omega)\, d\omega = \sigma^2_{S''} = \sigma^2_{SF} + \sigma^2_{N_1F}$$

because $S(t)$ and $N_1(t)$ are independent. The values of $\sigma^2_{SF}$ and $\sigma^2_{N_1F}$ are finite even if they represent filtered white noise, provided $\int_0^\infty |f^*(\omega)|^2 d\omega$ is finite. In this case $\overline{I}(R : S'')$ is finite. See, for instance, Breiman (1969, p. 284).

## 12.6 Channel Capacity

### 12.6.1 Non-Optical Frequency Signals

Let $R(t) = S(t) + N(t)$ $(-\infty < t < \infty)$ where $S(t)$ and $N(t)$ are independent continuous-parameter or generalized random processes and where $N(t)$ is gaussian. Then it is always possible to find a gaussian signal $S_G$, independent of $N$, such that

$$I(R_T : S_T) \leq I(R_T : S_{G_T})$$

for all $T$. See, for instance, Balakrishnan (1968).

Therefore, the channel capacity, which is defined as

$$C = \sup_S \overline{I}(R : S)$$

for all possible input signals S is obtained by considering only gaussian signals.

If N(t) is white noise with constant spectral density K, then from expression (12.4.4.1) it follows that

$$C < \frac{1}{K} \int_0^\infty P^*_{S_G}(\omega)\, d\omega = \sigma^2_{S_G}/2K\,. \qquad (12.6.1.1)$$

This upper bound is not attained for any signal with finite variance.

Balakrishnan defines the channel capacity to be

$$C = \frac{1}{2} \int_\Omega \log \left(\frac{\sigma^2_{S_G} + \sigma^2_N}{M(\Omega)\; P^*_N(\omega)}\right) d\omega \qquad (12.6.1.2)$$

where $\Omega$ is a measurable subset of the real line, $P^*_N(\omega) \neq 0$ over $\Omega$, and $M(\Omega)$ is some measure over $\Omega$, which in most cases is Lebesgue measure.

This capacity may be achieved by using a signal $\hat{S}(t)$ such that $P^*_{\hat{S}}(\omega) = b - P^*_N(\omega)$ on an $\Omega_b$ chosen so that

$$\sigma^2_{\hat{S}} = \int_{\Omega_b} (b - P^*_N(\omega))\; d\omega$$

where $b > 0$. If N(t) is band-limited white gaussian noise between -W and W with constant spectral density K then

$$\sigma^2_N = \int_{-W}^{W} K\; d\omega = 2KW$$

so that $P^*_N(\omega) = K = \sigma^2_N/2W$ when $(-W < \omega < W)$. Therefore, from (12.6.1.2)

$$C = \frac{1}{2} \int_{-W}^{W} \log \frac{\sigma^2_{S_G} + \sigma^2_N}{2W\; (\sigma^2_N/2W)}\, d\omega = W \log\, (1 + \sigma^2_{S_G}/\sigma^2_N)$$

where $M(\Omega) = 2W$. This formula is identical with Shannon's (8.9.3.2) when S is also band-limited white noise between -W and W. Therefore, Equation (12.6.1.2) is only valid when sampling at the uncorrelated points. (See section 8.9.3.)

If, however, S is not assumed necessarily to be band-limited white noise then

$$\lim_{W\to\infty} C = \lim_{W\to\infty} \log\left[1 + \frac{\sigma^2_{S_G}}{2KW}\right]^W = \log e^{\sigma^2_{S_G}/2K} = \frac{\sigma^2_{S_G}}{2K}$$

which is equivalent to (12.6.1.1) and is infinite if $S_G$ is also white noise.

## 12.6.2 Optical Frequency Signals

Shannon's formula $C = W(1 + \frac{\sigma^2_S}{\sigma^2_N})$ is not valid when working with laser signals and other signals at optical frequencies even if the sample points are at the uncorrelated points. Instead the channel capacity is

$$C = W \log\left(1 + \frac{\sigma^2_S}{\sigma^2_N + hfW}\right) + \frac{\sigma^2_S + \sigma^2_N}{hf} \log\left(1 + \frac{hfW}{\sigma^2_S + \sigma^2_N}\right) - \frac{\sigma^2_N}{hf} \log\left(1 + \frac{hfW}{\sigma^2_N}\right)$$

where W is the system bandwidth, $\sigma^2_S$ is the variance of the gaussian signal, $\sigma^2_N$ is the variance of the gaussian noise, h is Plank's constant, and f is the radiation frequency. (See Ross (1966) and Liu (1970).)

Therefore the channel capacity in this case remains finite as the variance $\sigma^2_N$ of the thermal noise approaches zero.

## 12.7 Expected Mutual Information Rate for Multidimensional Random Processes

Let ${}_nX(t)' = (X_1(t), \ldots, X_n(t))$ $(-\infty < t < \infty)$ and ${}_mY(t)' = (Y_1(t), \ldots, Y_m(t))$ $(-\infty < t < \infty)$ represent continuous-parameter or generalized multidimensional strictly stationary random processes. Let ${}_nX' = \{{}_nX(t)\}$ $(-\infty < t < \infty)$ and ${}_mY' = \{{}_mY(t)\}$ $(-\infty < t < \infty)$. It should be noted that ${}_nX$ is not the same as

$X_n' = (X(t_1), \ldots, X(t_n))$. The expected mutual information rate between ${}_nX$ and ${}_mY$ is

$$\bar{I}({}_nX : {}_mY) = \lim_{T\to\infty} \frac{1}{T} I({}_nX_T : {}_mY_T)$$

where ${}_nX_T' = \{{}_nX(t)\}$ (t in T) and ${}_mY_T' = \{{}_mY(t)\}$ (t in T).

If ${}_nX(t_i)$ and ${}_mY(t_i)$ $(i = 1, 2, \ldots)$ represent discrete-parameter multidimensional strictly stationary random processes, then

$${}_nX = \{{}_nX(t_i)\} \; (i = 1, 2, \ldots) \text{ and}$$

$${}_mY = \{{}_mY(t_i)\} \; (i = 1, 2, \ldots)$$

and

$$\bar{I}({}_nX : {}_mY) = \lim_{\ell\to\infty} \frac{1}{\ell} I({}_nX_\ell : {}_mY_\ell)$$

where ${}_nX_\ell' = \{{}_nX(t_i)\}$ $(i = 1, 2, \ldots, \ell)$ and ${}_mY_\ell' = \{{}_mY(t_i)\}$ $(i = 1, 2, \ldots, \ell)$.

### 12.7.1 Multidimensional Stationary Gaussian Processes

The results of this section are from Pinsker (1964, p. 193). Define

$$\gamma_{X_iY_j}(\tau) = E[X_i(t)Y_j(t + \tau)]$$

$(i = 1, \ldots, n)$ $(j = 1, \ldots, m)$ for the above processes. Assume that $E[X_i(t)] = E[Y_j(t)] = 0$ $(i = 1, \ldots, n)$ $(j = 1, \ldots, m)$ and define

$$P^*_{X_iY_j}(\omega) = \int_{-\infty}^{\infty} \gamma_{X_iY_j}(\tau)\, e^{-i2\pi\tau\omega}\, d\tau \quad (-\infty < \omega < \infty)$$

to be the cross spectral density function. Define the matrices

$$A^*_X(\omega) = \left[P^*_{X_iX_j}(\omega)\right], \quad A^*_Y(\omega) = \left[P^*_{Y_iY_j}(\omega)\right], \quad A^*_{XY}(\omega) = \left[P^*_{X_iY_j}(\omega)\right]$$

and

$$A^*(\omega) = \begin{bmatrix} A^*_X(\omega) , & A^*_{XY}(\omega) \\ \\ A^*_{YX}(\omega) , & A^*_Y(\omega) \end{bmatrix} .$$

Let $|\tilde{A}^*_X(\omega)|$ and $|\tilde{A}^*_Y(\omega)|$ represent any highest order nonvanishing principal minors of $A^*_X(\omega)$ and $A^*_Y(\omega)$ , respectively. Let $|\tilde{A}^*(\omega)|$ represent the principal minor of $A^*(\omega)$ that contains $\tilde{A}^*_X(\omega)$ and $\tilde{A}^*_Y(\omega)$ (which are the matrices corresponding to the principal minors). Thus,

$$\tilde{A}^*(\omega) = \begin{bmatrix} \tilde{A}^*_X(\omega) , & \tilde{A}^*_{XY}(\omega) \\ \\ \tilde{A}^*_{YX}(\omega) , & \tilde{A}^*_Y(\omega) \end{bmatrix} .$$

When ${}_nX$ and ${}_mY$ are gaussian,

$$I({}_nX : {}_mY) = \int \log \frac{|\tilde{A}^*_X(\omega)| \; |\tilde{A}^*_Y(\omega)|}{|\tilde{A}^*(\omega)|} \, d\omega .$$

Notice the similarity between this formula and (4.6.0.4). We now consider some special cases.

(i) ${}_nX$ and ${}_mY$ are discrete-parameter stationary gaussian processes and each of the n components of ${}_nX$ or the m components of ${}_mY$ is a regular process. (See section 8.8.1 for the definition of "regular".) This agrees with Good (1960a, p. 133, A5.4). The limits of integration in this case are 0 and 0.5.

(ii) ${}_nX$ and ${}_mY$ are continuous-parameter or generalized stationary gaussian processes and the spectral and joint spectral density functions of the components of ${}_nX$ or ${}_mY$ are all rational functions of $\omega$. The limits of integration in this case are 0 and $\infty$.

If the spectral and joint spectral density functions of the components of ${}_nX$ and ${}_mY$ are all rational functions of $\omega$, then it follows that $\bar{I}({}_nX : {}_mY)$ is finite if and only if $I({}_nX_T : {}_mY_T)$ is finite for all $T > 0$.

Four further specializations will now be considered.

(i) If $n = m = 1$, then $\mathbf{A}^*(\omega)$ is $2 \times 2$ and

$$\overline{I}({}_nX : {}_mY) = \log \frac{P^*_X(\omega)\ P^*_Y(\omega)}{P^*_X(\omega)\ P^*_Y(\omega) - |P^*_{XY}(\omega)|^2}\, d\omega$$

which agrees with (12.2.1).

(ii) If $n = m$ and $E[X_i(t)Y_j(t)] = 0$ if $i \neq j$, then

$$\overline{I}({}_nX : {}_nY) = \sum_{j=1}^{n} \int \log \frac{P^*_{X_j}(\omega)\ P^*_{Y_j}(\omega)}{P^*_{X_j}(\omega)\ P^*_{Y_j}(\omega) - |P^*_{X_jY_j}(\omega)|^2}\, d\omega\ .$$

(iii) If $n = m$ and $R_j(t) = S_j(t) + N_j(t)$ $(-\infty < t < \infty)$ and $S_j$ and $N_j$ are independent for all $j$ $(j = 1, 2, \ldots, n)$ then

$$\overline{I}({}_nR : {}_nS) = \int \log \frac{|A^*_R(\omega)|}{|A^*_N(\omega)|}\, d\omega$$

provided that $|A^*_N(\omega)| \neq 0$ and $|A^*_R(\omega)| \neq 0$.

(iv) If $n = m$ and $R_j(t) = S_j(t) + N_j(t)$ as in (iii) and $E[R_i(t)\ R_j(t)] = 0$ if $i \neq j$ then

$$\overline{I}({}_nR : {}_nS) = \sum_{j=1}^{n} \int \log \left(1 + \frac{P^*_{S_j}(\omega)}{P^*_{N_j}(\omega)}\right) d\omega\ .$$

## XIII. RATE OF EXPECTED WEIGHT OF EVIDENCE

In this chapter the rate of expected weight of evidence is discussed with particular emphasis on stationary gaussian processes. The special case is investigated when the means are constant, but not necessarily zero. The rate of expected weight of evidence is then discussed for multidimensional random processes.

The means of all random processes in this chapter (except in Section 13.3) are assumed to be zero.

### 13.1 Introduction

Let $X_1 = \{X_1(t)\}$ $(-\infty < t < \infty)$ and $X_2 = \{X_2(t)\}$ $(-\infty < t < \infty)$ represent continuous-parameter or generalized strictly stationary random processes under two opposing hypotheses $H_1$ and $H_2$. Then the rate of expected weight of evidence is defined as

$$\overline{W}(H_1/H_2) = \lim_{T\to\infty} \frac{1}{T} W_T(H_1/H_2),$$

where $W_T(H_1/H_2)$ is defined in the same manner as in (9.1.1) to include white noise.

If $X_1 = \{X_1(t_i)\}$ $(i = 1, 2, \ldots)$ and $X_2 = \{X_2(t_i)\}$ $(i = 1, 2, \ldots)$ represent discrete-parameter strictly stationary random processes then

$$\overline{W}(H_1/H_2) = \lim_{n\to\infty} \frac{1}{n} W_n(H_1/H_2)$$

where $W_n(H_1/H_2)$ is the expected weight of evidence for the two processes $\{X_j(t_i)\}$ $(j = 1, 2)$ $(i = 1, 2, \ldots, n)$.

### 13.2 Stationary Gaussian Processes

The results of this section, including those given without proof, are from Pinsker (1964, pp. 198-201).

### 13.2.1 Discrete-Parameter Processes

Let $X_1$ and $X_2$ be discrete-parameter stationary gaussian processes and suppose $X_2$ is a regular process (defined in Section 8.8.1). Assume the ratio $P_1^*(\omega)/P_2^*(\omega)$ is bounded above. Then

$$\overline{W}(H_1/H_2) = \int_0^{0.5} \left[\frac{P_1^*(\omega)}{P_2^*(\omega)} - 1 - \log \frac{P_1^*(\omega)}{P_2^*(\omega)}\right] d\omega .$$

The above assumptions are always satisfied when $X_1 = N$ and $X_2 = S + N$ are regular.

### 13.2.2 Continuous-Parameter or Generalized Processes

Let $X_1$ and $X_2$ be continuous-parameter or generalized stationary gaussian processes, where $X_2$ has a rational spectral density function and is therefore regular. Then

$$\overline{W}(H_1/H_2) = \int_0^{\infty} \left[\frac{P_1^*(\omega)}{P_2^*(\omega)} - 1 - \log \frac{P_1^*(\omega)}{P_2^*(\omega)}\right] d\omega . \qquad (13.2.2.1)$$

If both $X_1$ and $X_2$ have rational spectral density functions and $P_{X_2}^*(\omega)$ is different from zero almost everywhere, then $\overline{W}(H_1/H_2)$ is infinite if and only if $W_T(H_1/H_2)$ is infinite for all T. The results of Section 10.4 follow from the above discussion.

In some cases it is possible that $\overline{W}(H_1/H_2) = \infty$ and $\overline{W}(H_2/H_1) < \infty$. It then follows from Section 7.8, for gaussian processes, that $P_{1T} \| P_{2T}$, and therefore $W_T(H_1/H_2)$, $W_T(H_2/H_1)$, and $J_T(1,2)$ are finite for all T. If, however, both $X_1$ and $X_2$ have rational spectral density functions, and $P_2(\omega)$ is different from zero almost everywhere, then $\overline{W}(H_1/H_2)$ is infinite if and only if $\overline{W}(H_2/H_1)$ is infinite, because $W_T(H_1/H_2)$ is infinite, and hence, $W_T(H_2/H_1)$ is infinite for all T.

## 13.2.3 $\overline{W}(H_{S+N}/H_N)$ for Continuous-Parameter or Generalized Processes

Let R = S + N where S and N are independent continuous-parameter or generalized stationary gaussian processes with zero means. If $P_N(\omega)$ is a rational function of $\omega$, then it follows from Equation (13.2.2.1) that

$$\overline{W}(H_{S+N}/H_N) = \int_0^\infty \left[ \frac{P_S^*(\omega)}{P_N^*(\omega)} - \log\left(1 + \frac{P_S^*(\omega)}{P_N^*(\omega)}\right) \right] d\omega \ .$$

If both $P_S^*(\omega)$ and $P_N^*(\omega)$ are rational functions of $\omega$, then

$$\overline{W}(H_{S+N}/H_N|H_N) = \lim_{T\to\infty} \frac{1}{T} \int_{\Omega_T} \log \frac{dP_{S_T+N_T}}{dP_{N_T}} \, dP_{N_T}$$

$$= -\overline{W}(H_N/H_{S+N}) = -\int_0^\infty \left[ \frac{P_N^*(\omega)}{P_S^*(\omega) + P_N^*(\omega)} - 1 - \log \frac{P_N^*(\omega)}{P_X^*(\omega)+P_N^*(\omega)} \right] d\omega \ .$$

$$= \int_0^\infty \left[ \frac{P_S^*(\omega)/P_N^*(\omega)}{1+(P_S^*(\omega)/P_N^*(\omega)} - \log\left(1 + \frac{P_S^*(\omega)}{P_N^*(\omega)}\right) \right] d\omega \ .$$

Therefore, it follows that

$$\overline{J}(S+N,N) = \lim_{T\to\infty} \frac{1}{T} J_T(1,2) = \overline{W}(H_{S+N}/H_N) + \overline{W}(H_N/H_{S+N})$$

$$= \int_0^\infty \left[ \frac{P_S^*(\omega)}{P_N^*(\omega)} - \frac{P_S^*(\omega)/P_N^*(\omega)}{1+P_S^*(\omega)/P_N^*(\omega)} \right] d\omega$$

$$= \int_0^\infty \frac{(P_S^*(\omega)/P_N^*(\omega))^2}{[1+(P_S^*(\omega)/P_N^*(\omega))]} \, d\omega \ .$$

These results are all similar to the results of Section 8.9.4 where $\alpha = P_S^*(\omega)/P_N^*(\omega)$ in which case $\overline{W}(H_{S+N}/H_N)$, $\overline{W}(H_{S+N}/H_N|H_N)$, $\overline{J}(S+N,N)$, $W_T(H_{S+N}/H_N)$,

$W_T(H_{S+N}/H_N|H_N)$ and $J_T(S+N,N)$ are infinite for all T. The first part of Section 11.3 follows from the above discussion.

## 13.2.4 Strictly Stationary Gaussian Processes with Constant Means not necessarily Zero

Let $X_1$ and $X_2$ be discrete or continuous-parameter or generalized strictly stationary gaussian processes with constant means $m_1$ and $m_2$, respectively, and spectral density functions $P_1^*(\omega)$ and $P_2^*(\omega)$, where $P_2^*(\omega)$ is a rational function of $\omega$. Then

$$\overline{W}(H_1/H_2) = \int \left[\frac{P_1^*(\omega)}{P_2^*(\omega)} - 1 - \log \frac{P_1^*(\omega)}{P_2^*(\omega)}\right] d\omega + \frac{(m_2-m_1)^2}{2\,P_2^*(0)} \qquad (13.2.4.1)$$

where the limits of integration are 0 and $\frac{1}{2}$ if the process is discrete, and 0 and $\infty$ in the continuous case.

For the special case when S is a non-random constant signal and N is a continuous or generalized stationary gaussian process with a rational spectral density function and a zero mean, then $P_N^*(\omega)$ and $P_{S+N}^*(\omega)$ are the same and $m_{S+N} = S$ and $m_M = 0$ so that the expression in (13.2.4.1) is

$$\overline{W}(H_N/H_{S+N}) = 0 + \frac{(S-0)^2}{2P_N^*(0)} = \frac{S^2}{2P_N^*(0)} \,. \qquad (13.2.4.2)$$

This is finite unless $P_N^*(0) = 0$. If $\int_{-\infty}^{\infty}|\gamma(\tau)|d\tau$ is finite then

$$P_N^*(0) = 2\int_0^{\infty} \gamma(\tau)d\tau \,.$$

Equation (13.2.4.2) does not apply when N is a singular process (see Section 8.8) such as a band-limited process, because $P_N^*(\omega)$ is not a rational function of $\omega$. However, if N is white noise, so that $P_N^*(\omega) = K$ $(-\infty < \omega < \infty)$ then $\overline{W}(H_N/H_{S+N}) = \frac{S^2}{2K} < \infty$.

## 13.3 Rate of Expected Weight of Evidence for Multidimensional Random Processes

Let

$$_n\mathbf{X}_1' = \{(X_{11}(t), \ldots, X_{1n}(t))\} \quad (-\infty < t < \infty)$$

and

$$_n\mathbf{X}_2' = \{(X_{21}(t), \ldots, X_{2n}(t))\} \quad (-\infty < t < \infty)$$

represent continuous-parameter or generalized multidimensional strictly stationary random processes under two opposing hypotheses $H_1$ and $H_2$. Then in the same manner as in Section 12.7 the expected weight of evidence is defined as

$$\overline{W}(H_1/H_2) = \lim_{T\to\infty} \frac{1}{T} W_T(H_1/H_2).$$

Let

$$_n\mathbf{X}_j = \{_nX_j(t_i)\} \quad (j = 1, 2; \; i = 1, 2, \ldots)$$

where

$$_nX_j(t_i)' = (X_{j1}(t_i), \ldots, X_{jn}(t_i)) \; .$$

$_n\mathbf{X}_j$ $(j = 1,2)$ represent discrete-parameter n-dimensional strictly stationary random processes under two opposing hypotheses $H_1$ and $H_2$. Then

$$\overline{W}(H_1/H_2) = \lim_{\ell\to\infty} \frac{1}{\ell} W_\ell(H_1/H_2)$$

where $W_\ell(H_1/H_2)$ is the expected weight of evidence in connection with the two processes $\{_nX_j(t_i)\}$ $(i = 1, \ldots, \ell)$ $(j = 1,2)$.

### 13.3.1 Multidimensional Stationary Gaussian Processes

The results of this section are from Pinsker (1964, pp. 196-200).

If $_n\mathbf{X}_1$ and $_n\mathbf{X}_2$ are two n-dimensional stationary gaussian processes then

$$\bar{W}(H_1/H_2) = \int \left[ \sum_{i=1}^{n} \sum_{k=1}^{n} P^*_{X_{1i}X_{1k}}(\omega) P^{*(-1)}_{X_{2i}X_{2k}}(\omega) - n - \log \frac{|\tilde{A}_{X_1}(\omega)|}{|\tilde{A}_{X_2}(\omega)|} \right] d\omega$$

holds for the cases (i) and (ii) below where $P^{*(-1)}_{X_{2i}X_{2k}}(\omega)$ represents a typical element of the inverse of the matrix $\tilde{A}^*_{X_2}(\omega)$ which is defined as in Section 12.7.1. The ranges of integration are 0 to $\frac{1}{2}$ in the discrete case, and 0 to $\infty$ in the continuous case.

(i) ${}_nX_2$ is a regular discrete-parameter random process and

$$\sum_{i=1}^{n} \sum_{k=1}^{n} P^*_{X_{1i}X_{1k}}(\omega)\ P^*_{X_{2i}X_{2k}}(\omega)$$

is bounded above. This will always be the case if ${}_nX_2 = {}_nN$ and ${}_nX_1 = {}_nS + {}_nN$ are independent.

(ii) ${}_nX_2$ is a regular continuous-parameter or generalized process with rational spectral and joint spectral density functions for all its components.

If both ${}_nX_1$ and ${}_nX_2$ are continuous-parameter or generalized processes with rational spectral and joint spectral density functions and each minor of the matrix $A_{X_2}(\omega)$ is either identically zero or almost everywhere different from zero then $\bar{W}(H_1/H_2)$ is infinite if and only if $W_T(H_1/H_2)$ is infinite for all T.

If the $X_{2i}(t)$ $(i = 1, 2, \ldots, n)$ are all mutually independent then

$$\bar{W}(H_1/H_2) = \sum_{i=1}^{n} \int \left[ \frac{P^*_{X_{1i}}(\omega)}{P^*_{X_{2i}}(\omega)} - 1 - \frac{1}{n} \log \frac{|\tilde{A}^*_{X_1}(\omega)|}{|\tilde{A}^*_{X_2}(\omega)|} \right] d\omega$$

which simplifies to the case of Section 13.2.1 or 13.2.2 when n = 1.

# XIV. GAUSSIAN PROCESSES WITH EQUAL COVARIANCE FUNCTIONS INCLUDING NONRANDOM SIGNALS IN GAUSSIAN NOISE

In this chapter gaussian processes with equal covariance functions, which include nonrandom signals in random noise, are discussed. Much of the discussion is related to the Karhunen-Loève representation. Necessary and sufficient conditions are given for the probability measures of two covariance-stationary gaussian processes to be parallel when their common spectral density function is rational. Several references are mentioned for further detail, such as for nongaussian processes.

Let $\mathbf{X}_1 = \{X_1(t)\}$ (t in T) and $\mathbf{X}_2 = \{X_2(t)\}$ (t in T) represent two gaussian processes with probability measures $P_{1T}$ and $P_{2T}$, respectively. Now suppose that $\gamma_1(t_1,t_2) = \gamma_2(t_1,t_2)$ where $\gamma_j(t_1,t_2)$ $(j = 1,2)$ are the covariance functions of $\mathbf{X}_1$ and $\mathbf{X}_2$ at any two points $t_1$ and $t_2$ in T.

Because gaussian processes are completely determined by their mean and covariance functions, $\mathbf{X}_1$ and $\mathbf{X}_2$ are identical unless their means are different. It may be assumed that $E[X_1(t)] = 0$, (t in T), since if it is not, the two processes may be replaced by $X_1(t) - E[X_1(t)]$ and $X_2(t) - E[X_1(t)]$.

Grenander (1950) found necessary and sufficient conditions for $P_{1T}$ and $P_{2T}$ to be parallel by using the Karhunen-Loève representation of the processes $\mathbf{X}_1$ and $\mathbf{X}_2$ with respect to eigenfunctions of the integral equation with a covariance kernel. See Section 14.1 for the definition of the Karhunen-Loève representation.

Parzen (1961) and Hájek (1962) found solutions for these conditions, as mentioned by Yaglom (1962, p. 331). These solutions are as follows.

Assume that $\mathbf{X}_1$ and $\mathbf{X}_2$ are covariance-stationary gaussian processes with fourier transforms $X_1^*(\omega)$ and $X_2^*(\omega)$ and common spectral distribution function

$F^*(\omega)$, and with means $m_1(t) = 0$ and $m_2(t)$ ($t$ in $T$). Then $P_{1T}$ and $P_{2T}$ are parallel or perpendicular depending on whether $m_2(t)$ can be represented in the form

$$m_2(t) = \int_{-\infty}^{\infty} e^{i2\pi t\omega} g^*(\omega)\, dF^*(\omega) \quad (t \text{ in } T) \tag{14.0.1}$$

where $g(\omega)$ is a complex function such that

$$\int_{-\infty}^{\infty} |g^*(\omega)|^2 dF^*(\omega)$$

is finite. If such a representation is possible then the Radon-Nikodym derivative is

$$\frac{dP_{2T}}{dP_{1T}} = \exp[-\int_{-\infty}^{\infty} g^*(\omega)\ X^*(\omega)\, d\omega + \frac{1}{2}\int_{-\infty}^{\infty} |g^*(\omega)|^2 dF^*(\omega)]\ .$$

## 14.1 The Karhunen-Loève Representation

Let $X = \{X(t)\}$ ($t$ in $T$) represent a stationary gaussian random process with mean zero. Let $\{\phi_i(t)\}$ be a set of orthonormal (not necessarily complete) functions satisfying

$$\int_0^T \gamma_X(s - t)\, \phi_i(t)\, dt = \lambda_i^2\ \phi_i(s) \quad (s \text{ in } T)$$

where $\gamma_X(s-t)$ is the covariance function of $X$. Then the Karhunen-Loève Representation Theorem states that there exists a sequence

$$X_i = \int_T X(t)\, \phi_i(t)\, dt \quad (i = 1, 2, \ldots)$$

where

$$E[X_i X_j] = \int_T \int_T \gamma(s - t)\, \phi_i(t)\, \phi_j(s)\, ds\ dt$$

$$= \lambda_i^2 \int_T \phi_i(s)\, \phi_j(s)\, ds = \lambda_i^2\ \delta_{ij}$$

such that

$$\lim_{n\to\infty} E[X(t) - \sum_{i=1}^{n} X_i \phi_i(t)]^2 = 0 \quad (t \text{ in } T).$$

As pointed out in Section 9.2.4, $\gamma_X$ will always be positive definite if X has a spectral density function that is not identically zero so that $\lambda_i^2 > 0$ $(i = 1, 2, \ldots)$, and $\Phi = \{\phi_i\}$ will then be a complete orthonormal system in $L_2(T)$. [See Davenport and Root (1958, pp. 340, 373).]

## 14.2 Nonrandom Signals in Random Noise

Let $S = \{S(t)\}$ (t in T) represent a non-random signal and let $N = \{N(t)\}$ (t in T) represent random gaussian noise with mean zero. Then $\gamma_{S+N}(t_1,t_2) = \gamma_N(t_1,t_2)$ so that this is a special case of the processes defined in the introduction to this chapter where $X_1 = N$ and $X_2 = S + N$.

When S is a constant and N is stationary with a rational spectral density function $P_N^*(\omega)$, then $\overline{W}(H_N/H_{S+N}) = \dfrac{S^2}{2P_N^*(0)}$, as pointed out in Section 13.2.4 and will be finite provided $P_N(0) \neq 0$.

### 14.2.1 Root

Davenport and Root (1958), Kelley, Reed and Root (1960), Root (1962), and Root (1968) expand the work of Grenander (1950), which makes use of the Karhunen-Loève representation described in Section 14.1. They extend it to complex processes, and give practical considerations that prevent perfect non-random signal detection in gaussian noise.

### 14.2.2 Evaluation of the Karhunen-Loève Representation

Kadota (1964), Kailath (1967), and Davenport and Root (1958, Art. 6-4), point out some advantages and disadvantages of using the Karhunen-Loève representation.

The advantages over other Fourier and sampling expansions is that the coefficients $X_i$ (see Section 14.1) are independent, so that direct application of suitable convergence theorems for independent random variables may be used.

The disadvantages are that the eigenvalues $\lambda_i$ and the eigenfunctions $\phi_i$ are difficult to compute, and do not have much physical meaning.

### 14.2.3 Middleton and Kailath

Middleton (1960, p. 728) gives a few solutions to eigenvalues, in terms of trigonometrical and Bessel functions, for the following integral equation

$$\int A(u)K(|t - u|)\ \phi_j(u)du = \lambda_j\phi_j(t)\ .$$

These are helpful for particular applications of the Karhunen-Loève expansion. The condition $\Sigma X_1^2/\lambda_i = \infty$ is necessary and sufficient for the singular detection problem, where $X_i$ and $\lambda_i^2$ are defined in Section 14.1 (Kailath, 1966a, p. 132). Kailath (1966b) also gives several other references for solving integral equations with covariance kernels.

### 14.2.4 Parzen

Parzen (1962) shows that $P_{N_T}$ and $P_{S+N_T}$ are parallel if and only if the equation (15.3.6.1) is satisfied when S is a non-random signal and N represents gaussian noise with mean zero. He shows this by using the theory of reproducing Hilbert spaces described in Section 15.3.6.

He also shows that $P_{N_T}$ and $P_{S_T+N_T}$ are equivalent if and only if S, which is the mean of S + N, satisfies equation (14.0.1) where $X_2 = S + N$ and $X_1 = N$.

### 14.2.5 Capon

Capon (1964) generalizes Parzen's results, but his results are not as easy to use as Parzen's. He points out that in reproducing Hilbert spaces, convergence in mean square implies pointwise convergence.

### 14.2.6 Martel and Matthews

Martel and Matthews (1961) give an example of perfect detection of non-random signals in band-limited gaussian noise and show the rate of approach to perfect detection as the sample size approaches infinity. [See also Gaarder (1966) and (1967), and Boverie and Gregg (1971).]

### 14.2.7 Selin

Selin (1965a) discusses three types of singularity between $P_{S_T+N_T}$ and $P_{N_T}$.

(i) Singularity when $T \geq T_0$, but non-singularity when $T < T_0$.

(ii) Singularity over arbitrarily small intervals of time.

(iii) Singularity over infinite intervals, but non-singularity over all finite intervals.

### 14.2.8 Kailath

Kailath (1966a) gives some simple methods for establishing the possibility of singular detection of non-random signals in gaussian noise without using eigenvalues and integral equations. He also points out that singular detection is possible for two-dimensional random processes, as discussed by Gaarder (1966, 1967).

### 14.2.9 Pierre

Pierre (1969) discusses singularity for non-gaussian processes.

# XV. SUMMARY OF THE MAJOR RESULTS IN THIS SURVEY FOR GAUSSIAN PROCESSES, INCLUDING GAUSSIAN SIGNALS AND NOISE

This chapter summarizes the conditions resulting in perpendicularity and parallelism of probability measures of gaussian processes, with particular emphasis on gaussian signals and noise. It is mentioned in Section 15.3.1 that in most real communication problems white noise arises in some way.

## 15.1 Summary of the Conditions Resulting in Perpendicular Probability Measures of Gaussian Processes

In Section 8.7 it is shown that if $X_1(t)$ $(-\infty < t < \infty)$ is an ergodic strictly stationary random process with an analytic covariance function, such as any strictly stationary band-limited gaussian process with an absolutely continuous spectral distribution function, and $X_2(t)$ (t in T) is any other random process, then $P_{1T} \perp P_{2T}$, and $W_T(H_1/H_2)$, $\overline{W}(H_1/H_2)$, $J_T(1,2)$, and $\overline{J}(1,2)$ will all be infinite. For instance, when N(t) (t in T) is gaussian band-limited white noise then $P_{N_T} \perp P_{S_T+N_T}$, regardless of whether S(t) (t in T) is a random process, as shown in Section 8.7.1.2.

In Section 11.1 it is shown that if $R_T = \{S(t) + N(t)\}$ (t in T) where $S_T = \{S(t)\}$ (t in T) and $N_T = \{N(t)\}$ (t in T) are independent stationary gaussian processes with mean zero, then whenever $P_{N_T} \perp P_{S_T+N_T}$ it follows that $W_T(H_{S+N}/H_N)$, $J_T(S+N,N)$, and $I(R_T:S_T)$, and $\overline{W}(H_{S+N}/H_N)$, $\overline{J}(S+N,N)$, and $\overline{I}(R:S)$ will all be infinite.

As in Section 8.9.1 suppose that $X_1(t)$ (t in T) and $X_2(t)$ (t in T) are two gaussian processes with proportional covariance functions, where the multidimensional sample observations $\{X(t_i)\}$ ($t_i$ in T) (i = 1, ..., n) have non-singular covariance matrices for all integers n. Then $P_{1T} \perp P_{2T}$ and therefore, $W_T(H_1/H_2)$, $J_T(1,2)$, $\overline{W}(H_1/H_2)$, and $\overline{J}(1,2)$ are all infinite.

It is shown in Sections 8.9.2, 8.9.3, 8.9.4, 9.2.3.2, and 9.2.4 that whenever the signal and noise are independent gaussian stationary processes with mean zero and have proportional spectral density functions then $P_{S_T} \perp P_{S_T+N_T}$, and therefore, $W_T(H_{S+N}/H_N)$, $J_T(S+N,N)$, $I(\mathbf{R}_T:S_T)$, $\overline{W}(H_{S+N}/H_N)$, $\overline{J}(S+N,N)$, and $\overline{I}(\mathbf{R}:\mathbf{S})$ are all infinite. An example of this occurs when the signal and noise are both white noise or both band-limited white noise with the same band width.

In Sections 9.2.2.2, 9.2.5, and in the introductory section of Chapter X, conditions are given in terms of projections operators and integral operators with covariance kernels that will result in perpendicular probability measures and infinite values for the expected weight of evidence and the expected mutual information of gaussian processes.

In Section 11.3 it is shown that when the spectral density functions, $P_S^*(\omega)$ and $P_N^*(\omega)$ of stationary gaussian signals and noise, are rational functions of $\omega$, or are bounded and asymptotically inversely proportional to even powers of $\omega$, and $\lim_{|\omega|\to\infty} \frac{P_S^*(\omega)}{P_N^*(\omega)} \neq 0$, then $P_{S_T+N_T} \perp P_{N_T}$ and $W_T(H_{S+N}/H_N)$, $J_T(S+N,N)$, $I(\mathbf{R}_T:S_T)$, $\overline{W}(H_{S+N}/H_N)$, $\overline{J}(S+N,N)$, and $\overline{I}(\mathbf{R}:S)$ are all infinite.

## 15.2 Summary of the Conditions Resulting in Parallel Probability Measures of Gaussian Processes

In Sections 9.2.2.2 and 9.2.5 conditions were given, in terms of projection and integral operators, for $I(\mathbf{X}_T:Y_S)$ to be finite.

In Section 9.2.3.1, $I(\mathbf{R}_T:S_T)$ was shown to be finite when the noise was white and the signal was a stationary process with a continuous covariance function and a continuous spectral density function that is not identically zero.

In the introductory section of Chapter X, necessary and sufficient conditions are given in terms of integral operators with covariance kernels, for $W_T(H_1/H_2)$ to be finite.

In Section 11.3 it was pointed out that $I(R_T:S_T)$, $W_T(H_N/H_{S+N})$, and $J_T(N,S+N)$ are all finite provided the

$$\lim_{|\omega|\to\infty} \frac{P_S^*(\omega)}{P_N^*(\omega)} = 0$$

when the independent stationary gaussian signal and noise have rational spectral density functions $P_S^*(\omega)$ and $P_N^*(\omega)$, respectively.

In Section 11.1 it was shown that whenever $I(R_T:S_T)$ is finite then $W_T(H_N/H_{S+N})$ and $J_T(N,S+N)$ are finite when the signal and noise are independent gaussian processes.

In Section 12.2 it was mentioned that whenever $\overline{I}(X:Y)$ is finite then $I(X_T:Y_T)$ also is finite for all T.

In Section 12.4.4 it was shown that $\overline{I}(R:S)$ is finite if the gaussian noise is white and the stationary gaussian signal has finite variance. It therefore follows that $I(R_T:S_T)$, $W_T(H_N/H_{S+N})$, and $J_T(N,S+N)$ are also finite for all T.

In Section 12.4.5 it was shown that $\overline{I}(R:S)$ is finite if

$$\frac{P_S^*(\omega)}{P_N^*(\omega)} \sim a\,\omega^{-(1+\varepsilon)}$$

for large values of $\omega$ where $a > 0$, and $\varepsilon > 0$, and $P_S^*(\omega)$ and $P_N^*(\omega)$ are the spectral density functions of the stationary gaussian signal and noise, respectively.

In Section 12.5 where the stationary gaussian external noise and signal are passed through a time-invariant linear system which adds an additional stationary gaussian internal noise, then the expected mutual information rate is shown to be finite if the spectral density function $P_{N_2}^*(\omega)$ of the internal stationary gaussian noise satisfies

$$P^*_{N_2}(\omega) \sim \alpha\, \omega^{-(2-\varepsilon)}$$

for large $\omega$ where $\varepsilon > 0$, and if also the stationary gaussian signal has finite variance.

In Section 13.2.2 it was pointed out that if either $\overline{W}(H_1/H_2)$ or $\overline{W}(H_2/H_1)$ is finite, even though they may not both be finite, then $W_T(H_1/H_2)$ and $W_T(H_2/H_1)$ are both finite for all T when the stationary processes are gaussian.

In Section 13.2.4 it was shown that when S is a non-random constant signal and N is a continuous-parameter or generalized stationary gaussian process with a rational spectral density function and mean zero, then $\overline{W}(H_N/H_{S+N})$, $W_T(H_N/H_{S+N})$, and $J_T(H_N/H_{S+N})$ are all finite provided that the spectral density function of the noise, $P^*_N(\omega)$, does not vanish when $\omega = 0$.

## 15.3 Additional Considerations Resulting in Parallel Probability Measures of Gaussian Processes

### 15.3.1 Vainshtein and Zubakov

Vainshtein and Zubakov (1962), Appendix III, point out that in most communication problems white noise arises in some way in the form of thermal noise in resistors, or shot effect, or errors in reproducing, measuring, or processing input data, which will prevent $P_{S_T+N_T}$ and $P_{N_T}$ from being perpendicular, provided the signal has finite power (or variance). The white noise has constant spectral density in a sufficiently large frequency interval and falls off sufficiently slowly as $\omega$ approaches infinity.

### 15.3.2 Good

Good (1958, 1959, and 1960) has shown that if the signal and noise are periodic stationary gaussian band-limited white noise then the total amount of expected mutual information is finite provided the period is finite, but it may still be obtained by sampling at points in any arbitrarily small interval.

He also pointed out that the distance between sampling points is always greater than some positive constant $\varepsilon$ due to the inaccuracy of the measuring device, recording device, or human error, etc. Therefore, in practice, it would not be possible to obtain an infinite number of observations in an arbitrarily small interval, and hence, the expected mutual information should remain finite. He points out, however, that this does not completely resolve the problem since the expected mutual information should remain finite in a finite interval regardless of how accurately the data can be measured and recorded from the sample.

### 15.3.3 Swerling

Swerling (1960) points out that the expected mutual information will always be finite when the signal and noise are the result of passing a stationary gaussian signal with finite power and stationary gaussian white noise through a linear system with a band-limited impulse response function. (See Sections 8.2 and 8.2.1 for definitions.)

### 15.3.4 Root

Root (1962) points out that in practical situations there is always some form of white noise present. He then shows that when the signal and noise are independent stationary gaussian processes, and the signal has finite power and rational spectral density, then $P_{S_T+N_T}$ and $P_{N_T}$ will not be perpendicular.

### 15.3.5 Slepian

Slepian (1958), was the first to prove that perfect signal detection, when the signal and noise are independent gaussian stationary processes with rational spectral density functions, implies that $\lim \frac{P_S^*(\omega)}{P_N^*(\omega)} \neq 0$. (See Sections 10.4, 11.2, and 11.3.)

He points out that in practice this situation does not exist because measurements are not precise. However, this argument fails to explain why it should still not be possible to have perfect detection by taking precise measurements over any arbitrarily small interval.

15.3.6 Parzen

Parzen (1962) gives an elegant method of showing that $P_{S_T+N_T}$ and $P_{N_T}$ will be parallel if there exists a set of $c_i$ $(i = 1, 2, \ldots)$ such that

$$\lim_{n\to\infty} E[S(t) - \sum_{i=1}^{n} c_i \gamma_N(t_i,t)]^2 = 0 \qquad (15.3.6.1)$$

for all t in an interval T, so that S(t), t in T, is at least as smooth as the noise, where S(t) and N(t), t in T, are independent gaussian processes with mean zero and $\gamma_N(t_i,t) = E[N(t)N(t_i)]$.

He proves this by using the theory of reproducing Hilbert spaces $H(\gamma_X:T)$ where $\gamma_X(r,s)$ is the covariance function of a stochastic process X(t), t in T. Aronszajn (1950) introduced reproducing Hilbert spaces, which satisfy the following properties.

(i) The members of $H(\gamma_X:T)$ are real-valued functions on T.

(ii) For every t in T, $\gamma_X(\cdot,t)$ is in $H(\gamma_X:T)$.

(iii) For every t in T and function f in $H(\gamma_X:T)$, f(t) is equal to the inner product $(f(\cdot), \gamma_X(\cdot,t))$.

He shows that a sufficient condition for $P_{S_T+N_T}$ and $P_{N_T}$ to be parallel is that

$$\int_{-\infty}^{\infty} \frac{P_S^*(\omega)}{P_N^*(\omega)} d\omega < \infty ,$$

where $P_S^*(\omega)$ and $P_N^*(\omega)$ are the spectral density functions of the stationary gaussian processes S(t) and N(t) (t in T), respectively.

He proves this without using the information rate, as was done in expression (12.4.4.1). As mentioned in Section 12.4.5, Middleton (1961, p. 112) also arrived at this conclusion.

# XVI. CONCLUSIONS AND AREAS FOR ADDITIONAL RESEARCH

Several situations have been examined which result in the singularity between probability measures. This results in infinite values of the expected mutual information, expected weight of evidence, and divergence. Several conditions that will prevent these situations from occurring were also investigated. For instance, in communication theory, suppose one is discriminating between the hypothesis that the received signal R(t) is S(t) + N(t) (t in T) and the hypothesis that R(t) = N(t) (t in T) where N(t) is stationary gaussian white noise with mean zero, and the original signal S(t) (t in T) is either a stationary gaussian process, with zero mean, and finite variance, independent of the noise, or a non-random process. Then the probability measures defined on these processes under the two hypotheses were shown not to be perpendicular, but parallel, and therefore $W_T(H_{S+N}/H_N)$ and $J_T(S+N,N)$ are finite. In this case, perfect signal detection in additive noise does not occur over finite intervals. Also the expected information, $I(R_T:S_T)$, about the original gaussian signal, $S_T = \{S(t)\}$ (t in T), described above, provided by the received signal, $R_T = \{S(t) + N(t)\}$ (t in T) is finite.

In the general case, however, it is not always easy to tell whether a probability measure defined under one hypothesis is singular with respect to a probability measure under another hypothesis. It is sometimes difficult to tell whether the conditions given for the singularity between gaussian probability measures are satisfied.

There is much room for additional research on these problems. Much of the functional analysis as well as the setting up of realistic models in communication theory needs to be developed further. There has been very little relevant work done for nongaussian and nonstationary processes.

## BIBLIOGRAPHY

The items marked with an asterisk came to our attention after the manuscript was completed. Apart from these there are only a few other items that we have not seen.

Alekseev, V. G. (1963) "On conditions for perpendicularity of Gaussian measures corresponding to two stochastic processes", *Theor. Probability Appl.*, **8**, 286-290.

Anderson, T. W. (1971) *The Statistical Analysis of Time Series*, Wiley, New York.

Apokorin, D. S. (1967) "Gaussian measures corresponding to generalized stationary processes", *Theor. Probability Appl.*, **12**, 638-646.

Aronszajn, N. (1950) "Theory of reproducing kernels", *Trans. Amer. Math. Soc.*, **68**, 337-404.

Baker, C. R. (1969a) "Mutual information for Gaussian processes". Paper presented at the *International Symposium on Information Theory* at Ellenville, New York, January, 1969. Also in *SIAM J. Appl. Math.*, **19** (1970), 451-458.

Baker, C. R. (1969b) "On the deflection of a quadratic-linear test statistic", *IEEE Trans. Information Theory*, **IT-15**, 16-21.

Baker, C. R. (1969c) "Complete simultaneous reduction of covariance operators", *SIAM J. Appl. Math.*, **17**, 972-983.*

Baker, C. R. (1971) "Detection and Information Theory", *Proc. Southeastern Symposium on System Theory*, pp. H-3 to H-3-11 (Invited Paper).*

Baker, C. R. (1973) "On equivalence of probability measures", *Annals of Probability*, 1, 690-698.*

Baker, C. R. (1974) "Zero-one laws for Gaussian measures on Banach space", *Trans. American Math. Society* (to appear, Jan. 1974).*

Baker, C. R. (to appear) "Joint measures and cross-covariance operators", *Trans. American Math. Society*.*

Balakrishnan, A. V. (Ed.) (1968) *Communication Theory*, McGraw-Hill, New York.

Bartlett, M. S. (1966) *An Introduction to Stochastic Processes*, 2nd Ed., Cambridge Univ. Press, London and New York.

Battail, G. (1964) "On the use of band-limited signals in communication theory", *Ann. Télécommun.*, 19, 125-137.

Bello, P. (1961) "Some results on the problem of discriminating between two Gaussian processes", *IEEE Trans. Information Theory*, **IT-7**, 224-233.

Belyaev, Yu K. (1959) "Analytic random processes", *Theor. Probability Appl.*, **4**, 402-409.

Boverie, B. and Gregg, W. D. (1971) "A signal detectability approach to optimization of the geometry of distributed aperture (array) receivers", *IEEE Trans. Antennas and Propagation*, **AP-19**, 22-30.

Breiman, L. (1969) *Probability and Stochastic Processes*, Houghton Mifflin Co., Boston, Mass.

Brillouin, L. (1956) *Science and Information Theory*, Academic Press, New York.

Cameron, R. H. and Martin, W. T. (1947) "The behaviour of measure and measurability under change of scale in Wiener space", *Bull. Amer. Math. Soc.*, **53**, 130-137.

Capon, J. (1964) "Radon-Nikodym derivative of stationary Gaussian measures", *Ann. Math. Statist.*, **35**, 517-531.

Condon, E. U. and Odishaw, H. (Eds.) (1958) *Handbook of Physics*, McGraw-Hill, New York.

Courant, R. and Hilbert, D. (1953) *Methods of Mathematical Physics*, **1**, Interscience, New York and London.

Cramér, H. (1937) *Random Variables and Probability Distributions*, Cambridge Univ. Press, London and New York.

Cramér, H. (1946) *Mathematical Methods of Statistics*, Univ. Press, Princeton, New Jersey.

Crum, M. M. and Good, I. J. (1962) Private Correspondence.

Davenport, W. B. and Root, W. L. (1958) *An Introduction to the Theory of Random Signals and Noise*, McGraw-Hill, New York.

Davis, R. (1954) "On the detection of sure signals in noise", *J. Appl. Phys.*, **25**, 76-82.

Dobrushin, R. L. (1961) "Mathematical problems in the Shannon theory of optimal coding of information", *Proc. 4th Berkeley Symposium on Mathematical Statistics and Probability*, *Vol.* ***1***, 211-252.

Dobrushin, R. L. (1963) "General formulation of Shannon's main theorem in information theory", *Amer. Math. Soc. Transl.*, ser. 2, **33**, 323-438. (Originally published in Russian in 1959.)

Dobrushin, R. L. (1972) "Survey of Soviet research in information theory", *IEEE Trans. Information Theory*, **IT-18**, 703-724. (Translated by E. R. Berlekam.) (254 references.) Covers 1953 to 1969.*

Doob, J. L. (1953) *Stochastic Processes*, Wiley, New York.

Duncan, T. (1970a) "Likelihood functions for stochastic signals in white noise", *Information and Control*, **16**, 303-310.

Duncan, T. (1970b) "On the absolute continuity of measures", *Ann. Math. Statist.*, **41**, 30-38.

Duttweiler, D. and Kailath, T. (1973) "RKHS approach to detection and estimation problems - Part IV: non-Gaussian detection", *IEEE Trans. Information Theory*, **IT-19**, 19-28.*

Duttweiler, D. and Kailath, T. (1973) "RKHS approach to detection and estimation problems - Part V: parameter estimation", *IEEE Trans. Information Theory*, IT-**19**, 29-37.*

Fang, Geng-Seng (1972) "Signal classification through quasi-singular detection with applications in mechanical fault diagnosis", *IEEE Trans. Information Theory*, **IT-18**, 631-636.*

Fano, R. M (1961) *Transmission of Information*, MIT Press and Wiley, New York.

Feldman, J. (1958) "Equivalence and perpendicularity of Gaussian processes", *Pacific J. Math.*, **8**, 699-708 and correction **9** (1959), 1295-1296.

Franks, L. (1969) *Signal Theory*, Prentice Hall, Englewood Cliffs, New Jersey.

Gaarder, N. T. (1966) "The design of point detector arrays, II", *IEEE Trans. Information Theory*, **IT-12**, 112-120.

Gaarder, N. T. (1967) "The design of point detector arrays, I", *IEEE Trans. Information Theory*, IT-13, 42-50.

Gallager, R. G. (1968) *Information and Reliable Communication*, Wiley, New York.

Gelfand, I. M. and Yaglom, A. M. (1959) "Calculation of the amount of information about a random function contained in another such function", *Amer. Math. Soc. Transl.*, ser. 2, **12**, 199-246. (Original version published in 1957.)

Gladyshev, E. G. (1961) "A new limit theorem for random processes with Gaussian increments", *Theor. Probability Appl.*, **6**, 52-61.

Golosov, J. (1966) "Gaussian measures equivalent to Gaussian Markov measures", *Soviet Mathematics*, **7**, 48-52.

Good, I. J. (1950) *Probability and the Weighing of Evidence*, Griffin, London and Hafner Publ., New York.

Good, I. J. (1956) "Some terminology and notation of information theory", *Proc. Inst. Elec. Engrs. C*, **103**, 200-204 [or Monograph No. 155R (1955)].

Good, I. J. (1960a) "Effective sampling rates for signal detection: or can the Gaussian model be salvaged?", *Information and Control*, **3**, 116-140.

Good, I. J. (1960b) "Weight of evidence, corrobation, explanatory power, information, and the utility of experiments", *J. Roy. Statist. Soc. Ser. B*, **22**, 319-331; Corrigenda, **30** (1968), 203.

Good, I. J. (1961) "Weight of evidence, causality, and false alarm probabilities", *Fourth London Symposium on Information Theory* (ed. Colin Cherry, London: Butterworths), 125-136.

Good, I. J. (1963a) "Information theory: survey", Communications Research Division - Institute for Defense Analysis, Working Paper No. 83, April, pp. 33.

Good, I. J. (1963b) "Maximum entropy for hypothesis formulation, especially for multidimensional contingency tables", *Ann. Math. Statist.*, **34**, 911-934.

Good, I. J. (1966) "A derivation of the probabilistic explication of information", *J. Roy. Statist. Soc. Ser. B*, **28**, 578-581.

Good, I. J. (1967) "On the principle of total evidence", *British J. Philos. Sci.*, **17**, 319-322.

Good, I. J. (1969a) "What is the use of a distribution?", in *Second International Symposium on Multivariate Analysis, Vol. 2* (ed. P. R. Krishnaiah Academic Press Inc., New York), 183-203.

Good, I. J. (1969b) "A subjective evaluation of Bode's law and an 'objective' test for approximate numerical rationality", *J. Amer. Statist. Assoc.*, **64**, 23-66 (with discussion).

Good, I. J. and Doog, K. Caj. (1958) "A paradox concerning rate of information", *Information and Control*, **1**, 113-126; **2** (1959), 195-197.

Good, I. J. and Toulmin, G. H. (1968) "Coding theorems and weight of evidence", *J. Inst. Math. Appl.*, **4**, 94-105.

Granger, C. W. and Hatanaka, M. (1964) *Spectral Analysis of Economic Time Series*, Princeton Press, Princeton, New Jersey.

Grenander, U. (1950) "Stochastic processes and statistical inference", *Ark. Mat.*, **1**, 195-277.

Hájek, J. (1958a) "A property of J-divergences of marginal probability distributions" (in English) *Czechoslovak Math. J.*, **8**(83), 460-463.

Hájek, J. (1958b) "On a property of normal distributions of any stochastic process", *Czechoslovak Math. J.*, **8**(83), 610-617. English translation in *Selected Translations in Mathematical Statistics and Probability*, **1**, Providence, R.I., 1961, 245-252.

Hájek, J. (1962) "On linear statistical problems in stochastic processes", *Czechoslovak Math. J.*, **12**(87), 404-444.

Hájek, J. (1964) "Strictly equivalent Gaussian measures" (Russian), *Proc. Fourth All-Union Math. Congr. (Leningrad 1961), Vol. II, Izdat. "Nauka"*, Leningrad, 1964, 341-345. English translation in *Selected Translations in Mathematical Statistics and Probability*, 9 (1971), 25-30.*

Helstrom, C. W. (1965) "Quantum limitations on the detection of coherent and incoherent signals", *IEEE Trans. Information Theory*, **IT-11**, 482-490.

Helstrom, C. W. (1968a) *Statistical Theory of Signal Detection*, 2nd Ed., Oxford: Pergamon.

Helstrom, C. W. (1968b) "Markov processes and applications", *Communication Theory* (ed. A. V. Balakrishnan, McGraw-Hill, New York), 26-87.

Hitsudu, M. (1968) "Representations of Gaussian processes equivalent to Wiener process", *Osaka J. Math.*, **5**, 299-312.

Huang, R. Y. and Johnson, R. A. (1962) "Information capacity of time-continuous channels", *IEEE Trans. Information Theory*, **IT-8**, S191-S198.

Huang, R. Y. and Johnson, R. A. (1963) "Information transmission with time-continuous random processes", *IEEE Trans. Information Theory*, **IT-9**, 84-94.

Jeffreys, H. (1946) "An invariant form for the prior probability in estimation problems", *Proc. Roy. Soc. London, Ser. A,* **186**, 453-461.

Jeffreys, H. (1948) *Theory of Probability,* 2nd Ed., Oxford Univ. Press (Clarendon), London and New York.

Jelinek, F. (1968) *Probabilistic Information Theory: Discrete and Memoryless Models,* McGraw-Hill, New York.

Jenkins, G. M. and Watts, D. G. (1968) *Spectral Analysis and its Applications,* Holden-Day, San Francisco.

Jørsboe, O. G. (1968) *Equivalence or Singularity of Gaussian Measures on Function Spaces,* Mathematisk Inst., Aarhus.

Kadota, T. T. (1964) "Optimum reception of binary Gaussian signals", *Bell System Tech. J.,* **43**, 2767-2810.

Kadota, T. T. (1965) "Optimum reception of binary sure and Gaussian signals", *Bell System Tech. J.,* **44**, 1621-1658.

Kadota, T. T. (1966) "Simultaneous orthogonal expansion of two stationary Gaussian processes - examples", *Bell System Tech. J.,* **45**, 1071-1096.

Kadota, T. T. (1967) "Differentiation of Karhunen-Loève expansion and application to optimum reception of sure signals in noise", *IEEE Trans. Information Theory,* **IT-13**, 255-260.

Kadota, T. T. (1970) "Nonsingular detection and likelihood ratio for random signals in white Gaussian noise", *IEEE Trans. Information Theory,* **IT-16**, 291-298.

Kadota, T. T. and Shepp, L. A. (1967) "On the best finite set of linear observables for discriminating two Gaussian signals", *IEEE Trans. Information Theory,* **IT-13**, 278-284.

Kadota, T. T. and Shepp, L. A. (1970) "Conditions for absolute continuity between a certain pair of probability measures", *Z. Wahrscheinlichkeitstheorie und Verw. Gebiete*, **16**, 250-260.*

Kadota, T. T.; Zakai, M.; and Ziv, J. (1971) "Mutual information of the white Gaussian channel with and without feedback", *IEEE Trans. Information Theory*, **IT-17**, 368-371.*

Kailath, T. (1966a) "Some results on singular detection", *Information and Control*, **9**, 130-152.

Kailath, T. (1966b) "Some integral equations with 'nonrational' kernels", *IEEE Trans. Information Theory*, **IT-12**, 442-447.

Kailath, T. (1971a) "RKHS [reproducing kernel Hilbert space] approach to detection and estimation problems - - Part I: deterministic signals in Gaussian noise", *IEEE Trans. Information Theory*, **IT-17**, 530-549.

Kailath, T. (1971b) "The structure of Radon-Nikodym derivatives with respect to Wiener and related measures", *Ann. Math. Statist.*, **42**, 1054-1067.*

Kailath, T. (1973) "An RKHS approach to detection and estimation problems - Part II: simultaenous diagonalization and Gaussian signals", submitted to *IEEE Trans. Information Theory*.*

Kailath, T. and Duttweiler, D. (1972) "An RKHS approach to detection and estimation problems - Part III: generalized innovations representations and a likelihood-ratio formula", *IEEE Trans. Information Theory*, **IT-18**, 730-745.*

Kailath, T. and Zakai, M. (1971) "Absolute continuity and Radon-Nikodym derivatives for certain measures relative to Wiener measure", *Ann. Math. Statist.*, **42**, 130-140.*

Kallianpur, G. and Mandekar, U. (1965) "Multiplicity and representation theory of purely non-deterministic stochastic processes", *Theory Prob. Applications,* **10**, 553-581.*

Kallianpur, G. and Oodaira, H. (1962) "The equivalence and singularity of Gaussian measures", *Proc. Symp. on Time-Series Analysis,* (ed. M. Rosenblatt, Wiley, New York), 279-291.

Kallianpur, G. and Oodaira, H. (1973) "Non-anticipative representations of equivalent Gaussian processes", *Annals of Probability,* **1**, 104-122.*

Kelley, E. J.; Reed, I. S.; and Root, W. (1960) "The detection of radar echoes in noise, Part I", *J. Soc. Ind. Appl. Math.,* **8**, 309-341.

Khinchin (Khintchine) A. Yu. (1934) "Korrelationstheorie der stationären stochastischen Prozesse", *Math. Ann.,* **109**, 604-615.

Khinchin, A. Yu. (1957) *Mathematical Foundations of Information Theory,* Dover Publ. Inc., New York.

Kolmogorov, A. N. (1941a) "Stationary sequences in Hilbert's space", *Bull. Math. Univ. Moscow,* **2**, No. 6, 40 pp. [English translation available at Courant Institute, New York University.]

Kolmogorov, A. N. (1941b) "Interpolation and extrapolation of stationary random sequences", *Bull. Acad. Sci. USSR, Math.,* **5**, 3-14 (Russian; and German summary). [An English translation has been published by the Rand Corp., Santa Monica, Calif. as Memo RM-3090-PR].

Kolmogorov, A. N. (1956) "On the Shannon theory of information transmission in the case of continuous signals", *IEEE Trans. Information Theory,* **IT-2**, 102-108.

Kolmogorov, A. N. (1963) "Theory of transmission of information", *Amer. Math. Soc. Transl.,* ser. 2, **33**, 291-321. (Originally publ. in Russian in 1959.)

Kullback, S. (1959) *Information Theory and Statistics*, Wiley, New York.

Kullback, S. and Leibler, R. (1951) "On information and sufficiency", *Ann. Math. Statist.*, **22**, 79-86.

Lathi, B. P. (1965) *Signals, Systems, and Communication*, Wiley, New York.

Linfoot, E. H. (1957) "An informational measure of correlation", *Information and Control*, **1**, 85-89.

Liu, J. W. (1970) "Reliability of quantum-mechanical communication systems", *IEEE Trans. Information Theory*, **IT-16**, 319-329.

Loève, M. (1963) *Probability Theory*, 3rd Ed., Van Nostrand, Princeton Univ. Press, Princeton, New Jersey.

Martel, H. C. and Mathews, M. V. (1961) "Further results on the detection of known signals in Gaussian noise", *Bell System Tech. J.*, **40**, 423-451.

Middleton, D. (1960) *An Introduction to Statistical Communication Theory*, International Ser. in Pure and Appl. Phys., McGraw-Hill, New York.

Middleton, D. (1961) "On singular and non-singular optimum (Bayes) tests for the detection of normal stochastic signals in normal noise", *IEEE Trans. Information Theory*, **IT-7**, 105-113 (see comments by Selin, I. and Middleton in **IT-8**, pp. 326 and 385-387).

Minsky, M. and Selfridge, O. (1961) "Learning in random nets" in *Information Theory: Fourth London Symposium* (ed. Colin Cherry; London: Butterworths), 335-347.

Mirskaja, T. I.; Pabedinskaite, A. S.; and Tempel'man, A. A. H. (1967) "Hilbert spaces of certain reproducing kernels and the equivalence of Gaussian measures", *Litovsk. Mat. Sb.*, **7**, 459-469. English translation in *Selected Transl. in Math. Statist. and Probability*, **11**, 121-131.*

Morrison, D. F. (1967) *Multivariate Statistical Methods*, McGraw-Hill, New York.

Munroe, M. E. (1953) *Introduction to Measure and Integration*, Addison-Wesley, Cambridge, Mass.

Newstead, G. (1959) *General Circuit Theory*, Wiley, New York.

Nyquist, H. (1928) "Certain topics in telegraph transmission theory", *Trans. Am. Inst. Elec. Engr.*, **47**, 617-644.

Pan, Yi-Min (1966) "Simple proofs of equivalence conditions for measures induced by Gaussian processes" (Chinese), *Shuxue Jinzhan*, **9**, 85-90. English translation in *Selected Transl. in Math. Statist. and Probability*, **12** (1973), 109-117.*

Papoulis, A. (1962) *The Fourier Integral and its Application*, McGraw-Hill, New York.

Papoulis, A. (1965) *Probability, Random Variables, and Stochastic Processes*, McGraw-Hill, New York.

Parzen, E. (1962a) "Probability density functionals and reproducing kernel Hilbert space", *Proc. Symp. on Time-Series Analysis* (ed. M. Rosenblatt, Wiley, New York), 155-169.

Parzen, E. (1962b) *Stochastic Processes*, Holden Day, San Francisco.

Parzen, E. (1967) *Time Series Analysis Papers*, Holden Day, San Francisco.

Peirce, C. S. (1878) "The probability of induction", *Popular Science Monthly*, reprinted in *The World of Mathematics, Vol. 2* (ed. James R. Newman; New York: Simon and Schuster, 1956), 1341-1354.

Pierce, J. R. (1973) "The early days of information theory", *IEEE Trans. Information Theory*, **IT-19**, 3-8.*

Pierre, P. A. (1969) "Singular non-Gaussian measures in detection and estimation theory", *IEEE Trans. Information Theory*, **IT-15**, 266-272.

Pierre, P. A. (1970a) "Asymptotic eigenfunctions of covariance kernels with rational spectral densities", *IEEE Trans. Information Theory*, **IT-16**, 346-347.

Pierre, P. A. (1970b) "On the covergence of error probabilities for signal detection", *Information and Control*, **17**, 10-13.

Pinsker, M. S. (1954) "The amount of information about a Gaussian random stationary process, contained in a second process", *Dokl. Akad. Nauk. SSSR*, **99**, no. 2, 213-216.

Pinsker, M. S. (1964) *Information and Information Stability of Random Variables and Processes* (English translation by A. Feinstein with notes) by Holden Day, San Francisco. (Originally published in Russian in 1960.)

Pisarenko, V. F. (1961) "Detection of random signals against a noise background", *Radio Engrg. Electron. Phys.*, **6**, 455-468.

Pitcher, T. S. (1966) "An integral expression for the log likelihood ratio of two Gaussian processes", *SIAM J. Appl. Math.*, **14**, 228-233.

Price, R. and Turin, G. L. (1963) "Communication and radar-section A", *IEEE Trans. Information Theory*, **IT-9**, 240-246. [Part of a Report on Progress in *Information Theory* in the U.S.A. by various authors.]

Rao, C. R. and Varadarajan, V. S. (1963) "Discrimination of Gaussian processes", *Sankhyā, Series A*, **25**, 303-330.

Rényi, A. (1967) "On some basic problems of statistics from the point of view of information theory", *Proc. 5th Berkeley Symposium on Mathematical Statistica and Probability, Vol. 1*, Berkeley and Los Angeles, Univ. California Press, 531-543.

Reza, F. M. (1961) *An Introduction to Information Theory*, McGraw-Hill, New York.

Riesz, F. and Nagy, B. Sz. (1956) *Functional Analysis*, London: Blackie.

Root, W. L. (1962) "Singular Gaussian measures in detection theory", *Proc. Symposium on Time-Series Analysis* (ed. M. Rosenblatt, Wiley, New York), 292-315.

Root, W. L. (1968) "The detection of signals in Gaussian noise", *Communication Theory* (ed. A. V. Balakrishnan, McGraw-Hill, New York), 160-191.

Root, W. L. and Pitcher, T. S. (1955) "Some remarks on statistical decisions", *IEEE Trans. Information Theory*, **IT-1**, No. 3, 33-38.

Ross, M. (1966) *Laser Receivers*, Wiley, New York.

Rozanov, Yu. A. (1962) "On the density of one Gaussian measure with respect to another", *Theor. Probability Appl.*, **7**, 82-87.

Rozanov, Yu. A. (1963) "On the problem of equivalence of probability measures corresponding to stationary Gaussian processes", *Theor. Probability Appl.*, **8**, 223-232.

Rozanov, Yu. A. (1964) "On probability measures in functional spaces corresponding to stationary Gaussian processes", *Theor. Probability Appl.*, **9**, 404-420.*

Rozanov, Yu. A. (1971) *Infinite-Dimensional Gaussian Distributions* (Providence, R.I.: Amer. Math. Soc.). Translation by G. Biruik from the Russian in *Proc. Steklov Inst. of Math. in the Acad. Sc. USSR*, **108** (1968).*

Sage, A. P. and Melsa, J. L. (1971) *Estimation Theory with Applications to Communications and Control*, McGraw-Hill, New York.

Sato, H. (1967) "On the equivalence of Gaussian measures", *J. Math. Soc. Japan*, **19**, 159-172.

Selin, I. (1962) "Comments on 'On singular and nonsingular optimum (Bayes) tests for detection of normal stochastic signals in normal noise'", by Middleton, D. (1961), *IEEE Trans. Information Theory*, **IT-8**, p. 326.

Selin, I. (1965a) "The sequential estimation and detection of signals in normal noise II", *Information and Control*, **8**, 1-35.

Selin, I. (1965b) *Detection Theory*, Princeton Univ. Press, Princeton, New Jersey.

Shannon, C. E. (1948) "A mathematical theory of communication", *Bell System Tech. J.*, **27**, 379-423 and 623-656. Reprinted by Univ. Ill. Press, Urbana, Ill., with an appendix by W. Weaver (1949a).

Shannon, C. E. (1949b) "Communication in the presence of noise", *Proc. I. R. E.*, **37**, 10-21.

Shepp, L. A. (1964) "The singularity of Gausian measures in function space", *Proc. Nat. Acad. Sci.*, **52**, 430-433.

Shepp, L. A. (1966) "Radon-Nikodym derivatives of Gaussian measures", *Ann. Math. Statist.*, **37**, 321-354.

Slater, J. C. (1939) *Introduction to Chemical Physics*, McGraw-Hill, New York.

Slepian, D. (1958) "Some comments on the detection of Gaussian signals in Gaussian noise", *IEEE Trans. Information Theory*, **IT-4**, 65-69.

Slepian, D. (1973) "Information theory in the fifties", *IEEE Trans. Information Theory*, **IT-19**, 145-148.*

Swerling, p. (1959) "Parameter estimation for waveforms in additive Gaussian noise", *J. Soc. Ind. Appl. Math.*, **7**, 152-166.

Swerling, p. (1960) "Paradoxes related to the rate of transmission of information", *Information and Control*, **3**, 351-359.

Swerling, p. (1966) "Detection of radar echoes in noise revisited", *IEEE Trans. Information Theory*, **IT-12**, 348-361.

Takahasi, H. (1965) "Information Theory of quantum-mechanical channels", *Advances in Communication Systems*, **1** (ed. A. V. Balakrishnan, Academic Press, New York), 227-310.

Tanner, W. P. (1960) "Theory of signal detectability as an interpretive tool for psychophysical data", *J. Acoust. Soc. Am.*, **32**, 1140-1147.

Thomas, J. B. (1969) *Introduction to Statistical Communication Theory*, Wiley, New York.

Turin, G. L. (1966) "Radar and communication theory: some unsolved problems", in *Progress in Radio Science, 1960-1963*, **6**; *Radio Waves and Circuits*, 194-199.

Turing, A. M. (1940) Personal communication with I. J. Good.

Van Trees, H. L. (1968) *Detection, Estimation and Modulation*, Wiley, New York.

Varberg, D. E. (1961) "On equivalence of Gaussian measures", *Pacific J. Math.*, **11**, 751-762.

Varberg, D. E. (1966) "On Gaussian measures equivalent to Wiener measure II", *Math. Scand.*, **18**, 143-160.

Viterbi, A. J. (1973) "Information theory in the sixties", *IEEE Trans. Information Theory*, **IT-19**, 257-262.*

Whittaker, E. T. (1915) "On the functions which are represented by the expansions of the interpolation-theory", *Proc. Roy. Soc. Edinburgh*, **35**, 181-194.

Wiener, N. (1930) "Generalized harmonic analysis", *Acta. Math.*, **55**, 117-258.

Wiener, N. (1949) *The Extrapolation, Interpolation, and Smoothing of Stationary Time Series with Engineering Applications*, MIT Press, Boston, Mass.

Woodward, P. M. (1953) *Probability and Information Theory with Applications to Radar*, Pergamon Press, London.

Yaglom, A. M. (1962) "On the equivalence and perpendicularity of two Gaussian probability measures in function space", *Proc. Symp. on Time-Series Analysis* (ed. M. Rosenblatt, Wiley, New York), 327-346.

Yaglom, A. M. (1963) "Mathematical theory of random functions", *Engrg. Cybernetics*, No. 5, 91-96. [Part X of a report on the Development of Information Theory in the USSR.]

Yagoda, I. (1969) "A note on optimum reception in Gaussian noise", *IEEE Trans. Information Theory*, **IT-15**, 322-323.

Yao, K. (1967) "Applications of reproducing kernel Hilbert space - band-limited signal models", *Information and Control*, **11**, 429-444.

Yeh, J. (1969) "Singularity of Gaussian measures in function spaces with factorable covariance functions", *Pacific J. Math.*, **31**, 547-554.

## SUBJECT INDEX

## AUTHOR INDEX

(The bibliography is not indexed.)

l. 215: P. Antonelli, D. Burghelea and P. J. Kahn, The Concordance-
›motopy Groups of Geometric Automorphism Groups. X, 140 pa-
›s. 1971. DM 16,–

l. 216: H. Maaß, Siegel's Modular Forms and Dirichlet Series.
, 328 pages. 1971. DM 20,–

l. 217: T. J. Jech, Lectures in Set Theory with Particular Emphasis on
e Method of Forcing. V, 137 pages. 1971. DM 16,–

l. 218: C. P. Schnorr, Zufälligkeit und Wahrscheinlichkeit. IV, 212
iten. 1971. DM 20,–

l. 219: N. L. Alling and N. Greenleaf, Foundations of the Theory of
ein Surfaces. IX, 117 pages. 1971. DM 16,–

l. 220: W. A. Coppel, Disconjugacy. V, 148 pages. 1971. DM 16,–

l. 221: P. Gabriel und F. Ulmer, Lokal präsentierbare Kategorien.
200 Seiten. 1971. DM 18,–

l. 222: C. Meghea, Compactification des Espaces Harmoniques.
108 pages. 1971. DM 16,–

l. 223: U. Felgner, Models of ZF-Set Theory. VI, 173 pages. 1971.
1 16,–

l. 224: Revètements Etales et Groupe Fondamental. (SGA 1). Dirigé
r A. Grothendieck XXII, 447 pages. 1971. DM 30,–

l. 225: Théorie des Intersections et Théorème de Riemann-Roch.
GA 6). Dirigé par P. Berthelot, A. Grothendieck et L. Illusie. XII,
0 pages. 1971. DM 40,–

l. 226: Seminar on Potential Theory, II. Edited by H. Bauer. IV,
0 pages. 1971. DM 18,–

l. 227: H. L. Montgomery, Topics in Multiplicative Number Theory.
178 pages. 1971. DM 18,–

l. 228: Conference on Applications of Numerical Analysis. Edited
J. Ll. Morris. X, 358 pages. 1971. DM 26,–

l. 229: J. Väisälä, Lectures on n-Dimensional Quasiconformal Map-
gs. XIV, 144 pages. 1971. DM 16,–

l. 230: L. Waelbroeck, Topological Vector Spaces and Algebras.
158 pages. 1971. DM 16,–

l. 231: H. Reiter, $L^1$-Algebras and Segal Algebras. XI, 113 pages.
71. DM 16,–

l. 232: T. H. Ganelius, Tauberian Remainder Theorems. VI, 75
ges. 1971. DM 16,–

l. 233: C. P. Tsokos and W. J. Padgett. Random Integral Equations
th Applications to stochastic Systems. VII, 174 pages. 1971. DM
–

. 234: A. Andreotti and W. Stoll. Analytic and Algebraic Depen-
nce of Meromorphic Functions. III, 390 pages. 1971. DM 26,–

l. 235: Global Differentiable Dynamics. Edited by O. Hájek, A. J.
hwater, and R. McCann. X, 140 pages. 1971. DM 16,–

. 236: M. Barr, P. A. Grillet, and D. H. van Osdol. Exact Cate-
ries and Categories of Sheaves. VII, 239 pages. 1971. DM 20,–

l. 237: B. Stenström, Rings and Modules of Quotients. VII, 136
ges. 1971. DM 16,–

. 238: Der kanonische Modul eines Cohen-Macaulay-Rings. Her-
sgegeben von Jürgen Herzog und Ernst Kunz. VI, 103 Seiten. 1971.
16,–

. 239: L. Illusie, Complexe Cotangent et Déformations I. XV, 355
ges. 1971. DM 26,–

. 240: A. Kerber, Representations of Permutation Groups I. VII,
2 pages. 1971. DM 18,–

. 241: S. Kaneyuki, Homogeneous Bounded Domains and Siegel
mains. V, 89 pages. 1971. DM 16,–

. 242: R. R. Coifman et G. Weiss, Analyse Harmonique Non-
mmutative sur Certains Espaces. V, 160 pages. 1971. DM 16,–

243: Japan-United States Seminar on Ordinary Differential and
nctional Equations. Edited by M. Urabe. VIII, 332 pages. 1971.
26,–

244: Séminaire Bourbaki – vol. 1970/71. Exposés 382–399.
356 pages. 1971. DM 26,–

245: D. E. Cohen, Groups of Cohomological Dimension One. V,
pages. 1972. DM 16,–

. 246: Lectures on Rings and Modules. Tulane University Ring
Operator Theory Year, 1970–1971. Volume I. X, 661 pages.
72. DM 40,–

Vol. 247: Lectures on Operator Algebras. Tulane University Ring and Operator Theory Year, 1970–1971. Volume II. XI, 786 pages. 1972. DM 40,–

Vol. 248: Lectures on the Applications of Sheaves to Ring Theory. Tulane University Ring and Operator Theory Year, 1970–1971. Volume III. VIII, 315 pages. 1971. DM 26,–

Vol. 249: Symposium on Algebraic Topology. Edited by P. J. Hilton. VII, 111 pages. 1971. DM 16,–

Vol. 250: B. Jónsson, Topics in Universal Algebra. VI, 220 pages. 1972. DM 20,–

Vol. 251: The Theory of Arithmetic Functions. Edited by A. A. Gioia and D. L. Goldsmith VI, 287 pages. 1972. DM 24,–

Vol. 252: D. A. Stone, Stratified Polyhedra. IX, 193 pages. 1972. DM 18,–

Vol. 253: V. Komkov, Optimal Control Theory for the Damping of Vibrations of Simple Elastic Systems. V, 240 pages. 1972. DM 20,–

Vol. 254: C. U. Jensen, Les Foncteurs Dérivés de $\varprojlim$ et leurs Applications en Théorie des Modules. V, 103 pages. 1972. DM 16,–

Vol. 255: Conference in Mathematical Logic – London '70. Edited by W. Hodges. VIII, 351 pages. 1972. DM 26,–

Vol. 256: C. A. Berenstein and M. A. Dostal, Analytically Uniform Spaces and their Applications to Convolution Equations. VII, 130 pages. 1972. DM 16,–

Vol. 257: R. B. Holmes, A Course on Optimization and Best Approximation. VIII, 233 pages. 1972. DM 20,–

Vol. 258: Séminaire de Probabilités VI. Edited by P. A. Meyer. VI, 253 pages. 1972. DM 22,–

Vol. 259: N. Moulis, Structures de Fredholm sur les Variétés Hilbertiennes. V, 123 pages. 1972. DM 16,–

Vol. 260: R. Godement and H. Jacquet, Zeta Functions of Simple Algebras. IX, 188 pages. 1972. DM 18,–

Vol. 261: A. Guichardet, Symmetric Hilbert Spaces and Related Topics. V, 197 pages. 1972. DM 18,–

Vol. 262: H. G. Zimmer, Computational Problems, Methods, and Results in Algebraic Number Theory. V, 103 pages. 1972. DM 16,–

Vol. 263: T. Parthasarathy, Selection Theorems and their Applications. VII, 101 pages. 1972. DM 16,–

Vol. 264: W. Messing, The Crystals Associated to Barsotti-Tate Groups: With Applications to Abelian Schemes. III, 190 pages. 1972. DM 18,–

Vol. 265: N. Saavedra Rivano, Catégories Tannakiennes. II, 418 pages. 1972. DM 26,–

Vol. 266: Conference on Harmonic Analysis. Edited by D. Gulick and R. L. Lipsman. VI, 323 pages. 1972. DM 24,–

Vol. 267: Numerische Lösung nichtlinearer partieller Differential- und Integro-Differentialgleichungen. Herausgegeben von R. Ansorge und W. Törnig, VI, 339 Seiten. 1972. DM 26,–

Vol. 268: C. G. Simader, On Dirichlet's Boundary Value Problem. IV, 238 pages. 1972. DM 20,–

Vol. 269: Théorie des Topos et Cohomologie Etale des Schémas. (SGA 4). Dirigé par M. Artin, A. Grothendieck et J. L. Verdier. XIX, 525 pages. 1972. DM 50,–

Vol. 270: Théorie des Topos et Cohomologie Etale des Schémas. Tome 2. (SGA 4). Dirigé par M. Artin, A. Grothendieck et J. L. Verdier. V, 418 pages. 1972. DM 50,–

Vol. 271: J. P. May, The Geometry of Iterated Loop Spaces. IX, 175 pages. 1972. DM 18,–

Vol. 272: K. R. Parthasarathy and K. Schmidt, Positive Definite Kernels, Continuous Tensor Products, and Central Limit Theorems of Probability Theory. VI, 107 pages. 1972. DM 16,–

Vol. 273: U. Seip, Kompakt erzeugte Vektorräume und Analysis. IX, 119 Seiten. 1972. DM 16,–

Vol. 274: Toposes, Algebraic Geometry and Logic. Edited by. F. W. Lawvere. VI, 189 pages. 1972. DM 18,–

Vol. 275: Séminaire Pierre Lelong (Analyse) Année 1970–1971. VI, 181 pages. 1972. DM 18,–

Vol. 276: A. Borel, Représentations de Groupes Localement Compacts. V, 98 pages. 1972. DM 16,–

Vol. 277: Séminaire Banach. Edité par C. Houzel. VII, 229 pages. 1972. DM 20,–

Vol. 278: H. Jacquet, Automorphic Forms on GL(2). Part II. XIII, 142 pages. 1972. DM 16,–

Vol. 279: R. Bott, S. Gitler and I. M. James, Lectures on Algebraic and Differential Topology. V, 174 pages. 1972. DM 18,–

Vol. 280: Conference on the Theory of Ordinary and Partial Differential Equations. Edited by W. N. Everitt and B. D. Sleeman. XV, 367 pages. 1972. DM 26,–

Vol. 281: Coherence in Categories. Edited by S. Mac Lane. VII, 235 pages. 1972. DM 20,–

Vol. 282: W. Klingenberg und P. Flaschel, Riemannsche Hilbertmannigfaltigkeiten. Periodische Geodätische. VII, 211 Seiten. 1972. DM 20,–

Vol. 283: L. Illusie, Complexe Cotangent et Déformations II. VII, 304 pages. 1972. DM 24,–

Vol. 284: P. A. Meyer, Martingales and Stochastic Integrals I. VI, 89 pages. 1972. DM 16,–

Vol. 285: P. de la Harpe, Classical Banach-Lie Algebras and Banach-Lie Groups of Operators in Hilbert Space. III, 160 pages. 1972. DM 16,–

Vol. 286: S. Murakami, On Automorphisms of Siegel Domains. V, 95 pages. 1972. DM 16,–

Vol. 287: Hyperfunctions and Pseudo-Differential Equations. Edited by H. Komatsu. VII, 529 pages. 1973. DM 36,–

Vol. 288: Groupes de Monodromie en Géométrie Algébrique. (SGA 7 I). Dirigé par A. Grothendieck. IX, 523 pages. 1972. DM 50,–

Vol. 289: B. Fuglede, Finely Harmonic Functions. III, 188. 1972. DM 18,–

Vol. 290: D. B. Zagier, Equivariant Pontrjagin Classes and Applications to Orbit Spaces. IX, 130 pages. 1972. DM 16,–

Vol. 291: P. Orlik, Seifert Manifolds. VIII, 155 pages. 1972. DM 16,–

Vol. 292: W. D. Wallis, A. P. Street and J. S. Wallis, Combinatorics: Room Squares, Sum-Free Sets, Hadamard Matrices. V, 508 pages. 1972. DM 50,–

Vol. 293: R. A. DeVore, The Approximation of Continuous Functions by Positive Linear Operators. VIII, 289 pages. 1972. DM 24,–

Vol. 294: Stability of Stochastic Dynamical Systems. Edited by R. F. Curtain. IX, 332 pages. 1972. DM 26,–

Vol. 295: C. Dellacherie, Ensembles Analytiques, Capacités, Mesures de Hausdorff. XII, 123 pages. 1972. DM 16,–

Vol. 296: Probability and Information Theory II. Edited by M. Behara, K. Krickeberg and J. Wolfowitz. V, 223 pages. 1973. DM 20,–

Vol. 297: J. Garnett, Analytic Capacity and Measure. IV, 138 pages. 1972. DM 16,–

Vol. 298: Proceedings of the Second Conference on Compact Transformation Groups. Part 1. XIII, 453 pages. 1972. DM 32,–

Vol. 299: Proceedings of the Second Conference on Compact Transformation Groups. Part 2. XIV, 327 pages. 1972. DM 26,–

Vol. 300: P. Eymard, Moyennes Invariantes et Représentations Unitaires. II. 113 pages. 1972. DM 16,–

Vol. 301: F. Pittnauer, Vorlesungen über asymptotische Reihen. VI, 186 Seiten. 1972. DM 18,–

Vol. 302: M. Demazure, Lectures on p-Divisible Groups. V, 98 pages. 1972. DM 16,–

Vol. 303: Graph Theory and Applications. Edited by Y. Alavi, D. R. Lick and A. T. White. IX, 329 pages. 1972. DM 26,–

Vol. 304: A. K. Bousfield and D. M. Kan, Homotopy Limits, Completions and Localizations. V, 348 pages. 1972. DM 26,–

Vol. 305: Théorie des Topos et Cohomologie Etale des Schémas. Tome 3. (SGA 4). Dirigé par M. Artin, A. Grothendieck et J. L. Verdier. VI, 640 pages. 1973. DM 50,–

Vol. 306: H. Luckhardt, Extensional Gödel Functional Interpretation. VI, 161 pages. 1973. DM 18,–

Vol. 307: J. L. Bretagnolle, S. D. Chatterji et P.-A. Meyer, Ecole d'été de Probabilités: Processus Stochastiques. VI, 198 pages. 1973. DM 20,–

Vol. 308: D. Knutson, λ-Rings and the Representation Theory of the Symmetric Group. IV, 203 pages. 1973. DM 20,–

Vol. 309: D. H. Sattinger, Topics in Stability and Bifurcation Theory. VI, 190 pages. 1973. DM 18,–

Vol. 310: B. Iversen, Generic Local Structure of the Morphisms i Commutative Algebra. IV, 108 pages. 1973. DM 16,–

Vol. 311: Conference on Commutative Algebra. Edited by J. W. Brew and E. A. Rutter. VII, 251 pages. 1973. DM 22,–

Vol. 312: Symposium on Ordinary Differential Equations. Edited b W. A. Harris, Jr. and Y. Sibuya. VIII, 204 pages. 1973. DM 22

Vol. 313: K. Jörgens and J. Weidmann, Spectral Properties of Ham tonian Operators. III, 140 pages. 1973. DM 16,–

Vol. 314: M. Deuring, Lectures on the Theory of Algebraic Function of One Variable. VI, 151 pages. 1973. DM 16,–

Vol. 315: K. Bichteler, Integration Theory (with Special Attention Vector Measures). VI, 357 pages. 1973. DM 26,–

Vol. 316: Symposium on Non-Well-Posed Problems and Logarithm Convexity. Edited by R. J. Knops. V, 176 pages. 1973. DM 18

Vol. 317: Séminaire Bourbaki – vol. 1971/72. Exposés 400–417. I 361 pages. 1973. DM 26,–

Vol. 318: Recent Advances in Topological Dynamics. Edited by Beck, VIII, 285 pages. 1973. DM 24,–

Vol. 319: Conference on Group Theory. Edited by R. W. Gatterda and K. W. Weston. V, 188 pages. 1973. DM 18,–

Vol. 320: Modular Functions of One Variable I. Edited by W. Kuyk. 195 pages. 1973. DM 18,–

Vol. 321: Séminaire de Probabilités VII. Edité par P. A. Meyer. VI, 32 pages. 1973. DM 26,–

Vol. 322: Nonlinear Problems in the Physical Sciences and Biolog Edited by I. Stakgold, D. D. Joseph and D. H. Sattinger. VIII, 3 pages. 1973. DM 26,–

Vol. 323: J. L. Lions, Perturbations Singulières dans les Problèm aux Limites et en Contrôle Optimal. XII, 645 pages. 1973. DM 42

Vol. 324: K. Kreith, Oscillation Theory. VI, 109 pages. 1973. DM 16

Vol. 325: Ch.-Ch. Chou, La Transformation de Fourier Complexe L'Equation de Convolution. IX, 137 pages. 1973. DM 16,–

Vol. 326: A. Robert, Elliptic Curves. VIII, 264 pages. 1973. DM 2

Vol. 327: E. Matlis, 1-Dimensional Cohen-Macaulay Rings. XII, 1 pages. 1973. DM 18,–

Vol. 328: J. R. Büchi and D. Siefkes, The Monadic Second Ord Theory of All Countable Ordinals. VI, 217 pages. 1973. DM 2

Vol. 329: W. Trebels, Multipliers for (C, α)-Bounded Fourier Expa sions in Banach Spaces and Approximation Theory. VII, 103 page 1973. DM 16,–

Vol. 330: Proceedings of the Second Japan-USSR Symposium Probability Theory. Edited by G. Maruyama and Yu. V. Prokhor VI, 550 pages. 1973. DM 36,–

Vol. 331: Summer School on Topological Vector Spaces. Edited L. Waelbroeck. VI, 226 pages. 1973. DM 20,–

Vol. 332: Séminaire Pierre Lelong (Analyse) Année 1971-1972. V, 1 pages. 1973. DM 16,–

Vol. 333: Numerische, insbesondere approximationstheoretische E handlung von Funktionalgleichungen. Herausgegeben von R. A sorge und W. Törnig. VI, 296 Seiten. 1973. DM 24,–

Vol. 334: F. Schweiger, The Metrical Theory of Jacobi-Perron Alg rithm. V, 111 pages. 1973. DM 16,–

Vol. 335: H. Huck, R. Roitzsch, U. Simon, W. Vortisch, R. Walden, Wegner und W. Wendland, Beweismethoden der Differentialgeom trie im Großen. IX, 159 Seiten. 1973. DM 18,–

Vol. 336: L'Analyse Harmonique dans le Domaine Complexe. Ed par E. J. Akutowicz. VIII, 169 pages. 1973. DM 18,–

Vol. 337: Cambridge Summer School in Mathematical Logic. Edit by A. R. D. Mathias and H. Rogers. IX, 660 pages. 1973. DM 4

Vol. 338: J. Lindenstrauss and L. Tzafriri, Classical Banach Spac IX, 243 pages. 1973. DM 22,–

Vol. 339: G. Kempf, F. Knudsen, D. Mumford and B. Saint-Don Toroidal Embeddings I. VIII, 209 pages. 1973. DM 20,–

Vol. 340: Groupes de Monodromie en Géométrie Algébriq (SGA 7 II). Par P. Deligne et N. Katz. X, 438 pages. 1973. DM 4

Vol. 341: Algebraic K-Theory I, Higher K-Theories. Edited H. Bass. XV, 335 pages. 1973. DM 26,–

Vol. 342: Algebraic K-Theory II, "Classical" Algebraic K-Theo and Connections with Arithmetic. Edited by H. Bass. XV, 5 pages. 1973. DM 36,–